A SHEARWATER BOOK

No Way Home ~

No Way Home

The Decline of the World's Great Animal Migrations

~

David S. Wilcove

with Illustrations by Louise Zemaitis

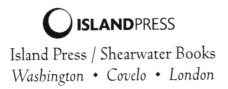

Island Press / Shearwater Books

Washington • *Covelo* • *London*

A *Shearwater Book*
Published by Island Press

SHEARWATER BOOKS is a trademark of
The Center for Resource Economics.

Library of Congress Cataloging-in-Publication data.
Wilcove, David Samuel.
No way home : the decline of the world's great animal migrations / by
David S. Wilcove.
p. cm.
Includes bibliographical references and index.
ISBN 978-1-55963-985-9 (hardcover : alk. paper)
1. Animal migration. 2. Animal populations. 3. Endangered ecosystems.
4. Environmental degradation. I. Title.
QL754.W55 2008
591.56'8—dc22
2007026205

British Cataloguing-in-Publication data available.
Portions of chapter 1 originally appeared in the January/February 1990
issue of *Nature Conservancy Magazine*.

Caption for frontispiece: Migration in the Serengeti: wildebeest and zebras

Printed on recycled, acid-free paper ⊕
Design by Joyce C. Weston
Illustrations by Louise Zemaitis, Swallowtail Studio
Manufactured in the United States of America

10 9 8 7 6 5 4 3 2

For Michael and Jonathan

Contents

On the Move

Every hour of every day, somewhere, some place, animals are on the move—flying, walking, crawling, swimming, or slithering from one destination to another. It is the ancient ritual of migration, and it is happening everywhere.

On a cold March morning, a Nebraska farmer pauses to admire a flock of sandhill cranes passing high overhead, their bugling call notes heralding the return of spring. On a hot, lazy September afternoon, a girl in Delaware chases after a monarch butterfly that is slowly but surely flapping its way to a mountaintop in central Mexico, where it will join tens of millions of its kin for the winter. In December, a Maasai tribesman climbs a rocky hillside and looks out across the savanna, where hundreds of thousands of wildebeest and zebras are heading from Kenya to Tanzania, tracking the ephemeral rain and the lush grass it promises to deliver, while on the other side of the world, along the shores of Laguna San Ignacio in Baja California, a fisherman awaits the return of the gray whales, which will use the bay's warm, shallow waters as a combination winter resort and nursery.

Cole Porter was right. Birds do it. Bees do it. Even bats with fleas do it.[1] Tens of thousands of species migrate, and the journeys they take are as different as the creatures themselves. Arctic terns migrate from their nesting grounds in the Bering Sea to the Antarctic Ocean, a circumpolar voyage that is without equal in the animal kingdom. At the other extreme, spotted salamanders in Maine awake from their winter hibernation in abandoned shrew burrows and trek 150 yards or so across the forest floor to their breeding ponds, an annual journey typical of many salamander species.[2] Three-wattled bellbirds in Costa Rica migrate from montane cloud forests to lowland jungles. Like the bellbirds, mountain quail in the western United States retreat from higher elevations during the winter, but they prefer to walk down the mountains rather than fly. Great white sharks will wander halfway across the Pacific Ocean and back over the course of a year, while krill, the little shrimplike crustaceans that are the bread and butter of the Southern Ocean's food chain, move up and down the water column in response to daylight. Theirs may be a daily migration of only a few hundred feet, but it is nonetheless essential for the survival of millions of other animals, ranging in size from two-ounce storm petrels to one-hundred-ton whales, that either consume krill or eat the creatures that consume krill.

Of course, animals are often on the move, and not all their wanderings fall into the category of migration. There is the daily search for food or shelter. There is the constant patrolling of one's territory to fend off intruders. And there is dispersal, which is movement away from a given site with no intention to return, as happens when young birds fledge from their nests and seek out their own territories or when fish larvae are carried away from their place of birth by ocean currents. None of these behaviors constitutes migration. Classic migration consists of seasonal back-and-forth journeys between two sites, as exemplified by the springtime reappearance of orioles in the backyard or alewives in the river.3 Typically, the travel occurs within a generation. In other words, the oriole that leaves New England in the fall will itself, if all goes well, return to New England in the spring. Some insects, however, spread their migrations out across generations. Monarch butterflies, for example, leave their wintering grounds in Mexico and fly north in the spring. Upon reaching the southeastern United States, they lay their eggs and die. The caterpillars hatch, develop into butterflies, and continue flying north. They, too, lay their eggs and die. This cycle is repeated for three or four generations until the butterflies have repopulated eastern North America as far north as New England and Canada. The generation that emerges in the late summer in the north then reverses course and heads south to Mexico.

Equally puzzling are the nomadic behaviors of certain birds, insects, and other animals that will depart from one location and wander for hundreds of miles in search of food. They follow no predictable course and appear to have no clear destination. Strictly speaking, such journeys fall somewhere between dispersal and migration. Yet given the distances these nomads travel, the varied habitats they visit, and the hardships they face, it makes little sense to exclude them from a discussion of the plight of migratory species. Accordingly, some are included in this book.

The means by which migratory animals navigate from place to place are as diverse as the journeys themselves. Some species follow an invisible road map created by the earth's magnetic field, which they perceive through tiny magnets in their bodies. Others rely on landmarks such as mountain ranges and coastlines, the alignment of the stars in the night sky, or olfactory cues to determine where they're going. Some even have a principle guidance mechanism and one or more backup systems—redundancy analogous to the back-up systems on commercial jets. Thus, on clear evenings, a migrating bird may navigate based on the apparent rotation of the stars, while on cloudy nights it can use the earth's magnetic field. For plenty of species, however, we simply don't know how they find their way. Yet somehow they manage to sniff, see, or sense when to go, where to go, and when to return.

At first, it's difficult to imagine any commonalities among migratory animals. The range of species involved, the different types of journeys they undertake, and the varying navigational tools they employ to reach their destinations defy easy categorization. Two traits, however, underlie most migratory phenomena. The first is opportunism. Migration enables animals to take advantage of abundant but ephemeral resources. The boreal forests of Canada, for example, are phenomenally rich in insects for a brief period of time, as any black-fly-slapping, mosquito-swatting hiker can attest. For an insect-eating bird, a summer in the North Woods can mean more than enough food to raise a large, healthy family, but only if the bird can get out before the onset of cold weather brings an end to the smorgasbord. Similarly, the Serengeti grasslands consist of little more than dirt and stubble for much of the year. But the arrival of the seasonal rains creates a flush of new growth, temporarily converting the parched landscape into a lush pasture well worth traveling dozens, even hundreds, of miles to visit, which is precisely why a million and a half wildebeest march there every year. Abundant, temporary food in all shapes and sizes is the lure for many migrants.

The second trait common to most migratory species is vulnerability, for even in the best of times the life of the migrant is no vacation. Storms, cold spells, heat spells, high winds, droughts, deluges, predators, parasites, diseases, and countless other dangers take their toll at every stage of the journey. Yet, despite the hardships, migration has proved to be a spectacularly successful strategy for numerous species. The sheer abundance of so many of them—the millions of wildebeest, zebras, and gazelles in East Africa; the hundreds of millions of warblers, vireos, and flycatchers in Canada and New England; the trillions of krill in the Antarctic Ocean—suggests that the journeys they make are well worth the effort.

However, that cost-benefit ratio seems to be changing in the face of a growing human population and its insatiable demand for natural resources. Simply stated, the phenomenon of migration is disappearing around the world. The great salmon runs of the Pacific Northwest have been reduced to a trickle as result of over a century of overexploitation, dam building, farming, livestock grazing, and logging. Monarch butterflies are threatened by illegal logging of the Mexican forests where they winter. In France, Italy, and Greece, the popular practice of shooting migrating songbirds as they cross the Mediterranean Sea has enraged conservationists and raised concern about the long-term health of Europe's bird populations. And birdwatchers across much of the eastern United States complain that the returning chorus of warblers, thrushes, and orioles grows fainter each spring, a consequence of forest destruction in the United States as well as in Latin America, where so many birds spend the winter. Migratory species after migratory species is in decline, regardless of where they live or how they travel.

The four great threats to migration (and, consequently, the four major actors in this book) are habitat destruction, human-created obstacles, overexploitation, and climate change. Habitat destruction is, of course, the best known menace to migration. Often described as the primary threat to wildlife across the globe, it strikes especially hard at migratory species because of the range of habitats they occupy during their travels. All migratory animals require a safe and secure destination at both ends of their journey. They also require safe and sufficient rest stops en route. Disrupt any part of the route—the breeding grounds, the wintering grounds, or the stopover sites in between—and the species is likely to suffer. Yet acre by acre we are transforming the world into a landscape wholly unsuited to many migratory species. Songbirds in search of winter quarters in the tropics increasingly encounter pastures and farm fields instead of forests; herds of elk and mule deer in the American West trek down from the mountains only to discover that the grasslands where they used to spend the winter have been turned into housing developments.

Obstructions come in as many shapes and sizes as the migratory species themselves. For salmon or shad, it can be a hydroelectric dam that blocks passage upstream. For aerial travelers like birds and bats, it may be a skyscraper or cell tower, replete with lights, that lures them to an untimely death. A simple barbed-wire fence stretched across a mountain pass may close off a migratory route that pronghorn antelopes have followed for millennia. And a newly constructed road that separates a pond from the adjoining uplands may be enough to eliminate a salamander population, one squashed salamander at a time. (I once spent a memorable night in Stanford, California, helping rare California tiger salamanders cross a busy road that separated their upland habitat from their breeding pond.) In a world of growing affluence and technological sophistication, more and more of the passageways for wildlife are being obstructed or destroyed.

Overexploitation may well be the oldest of the threats inasmuch as migratory animals have always been a crucial food source for people. For thousands of years, we hunted them with an array of crude but effective tools: spears, nets, traps, fire, fright (as when herds of bison were stampeded over cliffs by Plains Indians). Then, in the blink of an eye in evolutionary time, we upgraded to firearms and milewide drift nets; to refrigerated trains, trucks, boats, and airplanes that suddenly made it possible to catch a salmon off the coast of Washington and sell it the next day in Cleveland, Ohio; to a world where virtually any migratory species big enough to harvest can be harvested and then sold commercially; to a world where over six and a half billion people are clamoring for food. Small wonder, then, that those animals that aggregate in herds, schools, or flocks (as do so many migratory species) became especially sought-after targets.

As for climate change, it can best be described as the joker in the deck, its ultimate effect on migration as yet undetermined. Data gathered by birdwatchers, for example, suggest that some songbirds in Europe and North America are changing the timing of their migration in response to global warming, with spring migrants arriving earlier in the year. This would not necessarily be cause for concern if we could be certain that the resources these birds depend upon—insects, for example—will alter their emergence times in synch. But that doesn't appear to be the case, perhaps because the birds and bugs are responding to different cues. In other cases, critical habitats could literally disappear due to climate change. What will become of the sea turtles that have returned to certain beaches for generations to lay their eggs if those beaches are submerged by rising ocean levels? Will they seek out new nesting grounds, and, if so, where will they find them in an increasingly developed world? Migratory animals have weathered plenty of changes over the years, up to and including such dramatic events as the retreat of the Pleistocene ice sheets less than twelve thousand years ago. But the pace of change seems so much quicker today than in the past, leaving scientists to wonder which species will be able to cope and which will not.

Consider, for example, how much the world has changed for the chinook, sockeye, coho, chum, and pink salmon that annually swim up the Columbia River to spawn. No fewer than fourteen major dams along the Columbia and additional dams on the tributaries now block their progress. Fish ladders aid some of the adults as they head upstream but are almost useless to the juveniles heading downstream; the young fish are either stopped by the dams or ground into fishmeal by the turbines. Timber companies have stripped most of the primeval forest cover from the banks of the Columbia, sending tons of suffocating sediment into the water, while cattle and other livestock have consumed much of the stream-side vegetation and added their own unwelcome nutrients to the broth. Farmers in search of water to irrigate their crops have turned to the Columbia and its tributaries, resulting in lowered water levels, while offshore an expanding fleet of fishing boats hunts the remaining fish with radar, satellites, and massive nets. Small wonder, then, that the number of salmon making that journey today is barely a tenth of the number that did so two centuries ago.

Nor are the Columbia River salmon unique. All too many migratory animals face the same scenario: Bulldozers and plows are advancing on their breeding grounds and their wintering grounds; the journey back and forth has become increasingly arduous as stopover sites have been lost; if the migrants are hunted, it is by a growing army of people using increasingly sophisticated equipment; and the seasons themselves seem to be changing as an increasingly industrialized world pumps more and more greenhouse gases into the atmosphere.

Add to this the political hurdles conservationists face in attempting to protect migratory species, the primary one being geography. Migratory animals cross national borders with abandon. It's part of their appeal. But those same borders demarcate independent agencies, institutions, and cultures that somehow must coordinate their conservation efforts if the species is to prosper. In an increasingly fractious world, that's no small task. A lone Swainson's thrush traveling from its winter home in western Brazil to its breeding grounds in southern Manitoba will pass through (or over) ten different countries and more than forty states, provinces, departments, and other major subnational jurisdictions, not to mention hundreds of counties, municipalities, and towns. The fate of that

Barn Swallows

bird (and billions of other birds) rests with governments in Brazil, Colombia, Panama, Costa Rica, Nicaragua, Honduras, Guatemala, Mexico, the United States, and Canada—governments that have supported, invaded, aided, and competed with one another for decades or even centuries.

As daunting as the problem of coordinating conservation efforts across so many jurisdictions is, it may not be the biggest obstacle confronting people concerned about migratory species. A more fundamental problem may be that migration at its best is essentially a phenomenon of abundance. Just as one swallow does not a summer make, one warbler or one monarch does not constitute a migration—not, at least, in our hearts. We count on seeing not one, but thousands of warblers singing from the treetops in May or millions of monarchs winging their way southward in September. It is the sheer abundance of these animals that inspires and excites us. Plenty of parks in East Africa have wildebeest, but only the Serengeti has a million wildebeest that move in herds stretching from horizon to horizon.

Unfortunately, our whole approach to conservation has been reactive, rather than proactive: we wait until a species is in dire straits and then take steps to prevent its disappearance. Many migratory species therefore end up as low priorities for attention because they are still relatively common. Given how little money is available for conservation relative to other priorities, such as defense or health care, this type of triage makes a certain amount of sense. But one wonders how kindly future generations will judge us if they are deprived of even the diminished migrations we currently enjoy.

No country—rich or poor, strong or weak—has done an especially good job of protecting its transient wildlife. Indeed, to the people of the world's richest nation, the United States, falls the dubious distinction of having destroyed the two greatest migratory phenomena on earth: the bison of the Great Plains and the passenger pigeons of the eastern forests. At the start of the nineteenth century, perhaps as many as sixty million bison roamed the Great Plains.4 The ecology of the great bison herds will always be something of a mystery inasmuch as those who witnessed the herds were far more intent on destroying them than studying them. We do know the animals took part in a yearly migration from the plains, where they summered, to the river valleys and forested areas, where they wintered. In a frenzy of greed and blood lust lasting only a few decades, white settlers managed to eradicate the great herds and bring the bison to the edge of extinction by 1880.

The passenger pigeon was even less fortunate. At the start of the nineteenth century, the sleek, steel-blue doves nested by the tens of millions in immense aggregations scattered across the Northeast and Midwest. When the birds coalesced for their southward migration, the flocks became so large they were capable of obscuring the sun for days on end, casting an eerie twilight over the land. Ruthless overexploitation for commercial markets and for sport ultimately doomed the species. By the close of the nineteenth century, the passenger pigeon had vanished from the wild, and the last captive individual, an elderly female named Martha in the Cincinnati Zoo, died on September 1, 1914.

Fortunately, times and attitudes have changed, both in the United States and elsewhere around the world. If the end of migration has failed to attract the attention it deserves, it has nonetheless prompted some alarm among conservation-minded citizens and government agencies, giving rise to a growing number of national and international programs to protect migratory species. The question is whether those efforts will ever amount to enough to prevent the disappearance of still more migratory spectacles.

The irony is that just as the phenomenon of migration is slipping away, we are entering a golden age for studying it. Scientists recently discovered the wintering grounds of the endangered aquatic warbler by capturing a few of the birds in Europe, where they nest, and plucking a few feathers. Because the warblers arriving in Europe in the spring were in fresh plumage, the scientists knew the birds must have molted their old feathers and grown new ones while on the wintering grounds. By analyzing the chemical composition of the feathers, the scientists knew to focus on the region in West Africa bordering the Senegal River; a ground crew subsequently found the birds in northwest Senegal.[5] Transmitters weighing less than a dime can now be attached to creatures as small as a thrush, permitting scientists to follow them on their journey through the night skies. Much larger transmitters capable of conversing with satellites have already been placed on everything from great white sharks to polar bears to whooping cranes, enabling scientists to follow these animals around the clock from the comfort of their offices. It borders on the miraculous: a biologist, clad in blue jeans or khakis (as biologists usually are), stares at a computer screen as signals bounce back and forth from bear to satellite to computer in a minimigration of electromagnetic waves that reveals the real-life peregrinations of the animal.

In fact, almost every aspect of migration inspires awe: the incredible journeys migratory animals undertake and the hardships they face along the way; the complex mechanisms they use to navigate across the land and through the skies and seas; the sophisticated tools with which scientists study them; and, not least, the perseverance of the people striving to save these animals in the face of an increasingly congested, inhospitable world. It all adds up to one of the most daunting yet rewarding challenges in wildlife conservation.

In the Air~

CHAPTER 1

Empty Skies

Pick the right night and you will hear them. A September night, perhaps, when a northwest wind has swept the clouds from the sky and the stars are out in full force. As midnight approaches, find a quiet spot away from the rumbles and groans of urban life and listen carefully. Soon you will hear soft chirps and whistles drifting down from the sky. These are the calls of migrating songbirds. A thousand feet above you, extending for hundreds of miles in all directions, is a vast highway of little birds—millions of thrushes, warblers, flycatchers, tanagers, vireos, and sparrows—heading south. A few skilled birdwatchers can distinguish the call notes of the different species. A high, clear "chip" marks a northern parula warbler en route to the Caribbean; a sharp buzz identifies a black-poll warbler headed for the Amazon basin; a gentle whistle denotes a veery traveling from Canada to Colombia. The sky is alive.

The imagination takes flight. What would it be like to travel with the songbirds, to fly in the stillness of the night across the continents, illuminated from above by the stars and from below by the lights of cities? For the time being, the closest one can get to actually flying with the songbirds is to travel with Martin Wikelski. A professor of ecology at Princeton University, Wikelski is determined to figure out what he calls "the rules of migration," the physiological and behavioral cues birds use to decide when to migrate, where to go, and when to stop.

Wikelski's interest in birds dates back to his childhood in Bavaria, where as a teenager he captured and banded birds at a research station under the tutelage of local ornithologists. One spring, he decided to band the barn swallows that were nesting in his village. A year later, he recaptured some of the same birds in the same places. The realization that they had flown all the way to Africa and then returned to this tiny corner of Germany spawned a lifelong interest in migration. Blessed with a keen analytical mind, limitless energy, and an infectious enthusiasm for science, Wikelski is now one of the premier biologists of his generation. Dissatisfied with the idea of studying migratory birds in a laboratory, he instead has chosen to pursue them in the wild, taking advantage of recent breakthroughs in the field of electronics to track their movements.

Each spring for the past five years, Wikelski has journeyed to central Illinois to study migration.[1] Dawn on this particular day finds him stringing a line of nets through the middle of a small woodlot. The fineness of the mesh renders it almost invisible to birds, and before long an itinerant Swainson's thrush, flying through the trees, is ensnared. Wikelski gently extracts the tangled bird, smoothes its feathers, and weighs and measures it. He then attaches a radio transmitter, smaller than a fingernail and lighter than a dime, to the feathers on its back. He next injects the thrush with a tiny amount of doubly labeled water, which has a slightly different chemical signature than ordinary water but is otherwise identical and harmless. He then releases the bird, which darts into a nearby thicket and begins to preen its feathers fastidiously, as though trying to cleanse itself after the indignity of capture. Wikelski and coworkers retreat to the edge of the woodlot where, with the aid of a handheld antenna, they are now able to track the location of "their" thrush.

The bird spends most of the day searching for insect prey on the forest floor. Toward dusk, it appears to settle down for the night in a safe spot on the branch of a small tree. Motionless, eyes closed, it hardly seems like a bird on the go. But shortly before nine o'clock, the thrush stirs. It flies to another branch near the top of a taller tree and looks around. Suddenly, it launches into the night sky, compelled to complete its journey from its winter home in Brazil's lush rain forests to its summer home in the cool spruce forests of Canada. For Wikelski and company, the chase is on.

The scientists pile into an ancient wreck of a station wagon—a 1982 Ford—that has been customized for their research. A hole drilled through the roof allows them to insert a receiving antenna that can be controlled by the driver. The passenger's seat has been converted into a computer docking station, upon which sits a laptop linked to a GPS satellite. Wikelski clamps on a pair of headphones and listens for the signal from the transmitter on the migrating thrush. With his left hand on the steering wheel and his right hand manipulating the receiving antenna, he chases after the bird. The car careens down dark country roads at an alarming speed, the headlights illuminating row after row of young corn, in pursuit of the bird flying above them.

It's not a fair match. This thrush is built for air travel. Its hollow bones keep its body weight down to little more than an ounce. Its tapered wings are well designed to provide lift and forward velocity. As it cruises ahead at a ground speed of thirty-five miles per hour, a system of air sacs connected to its lungs assures a continuous, unidirectional flow of air through the lungs, greatly increasing the efficiency of each breath. Moreover, the bird is equipped with a remarkably sophisticated orientation and navigation system. It has an internal compass, capable of reading the earth's magnetic field, which it sets each day by noting the position of the setting sun. At night, it can determine its position and direction of travel by tracking the apparent movement of the stars, much the way sailors have done for millennia. The thrush, in short, instinctively knows where it wants to go and how to get there. The ornithologists, on the other hand, are restricted to the existing road network, which necessitates frequent midcourse corrections and an average speed well in excess of the thrush's. A colleague in the backseat tracks the thrush's journey on the computer and offers occasional guidance to Wikelski. "It's heading north-northeast now . . . turn right at the next intersection . . . drive faster, we're losing the signal."

They will drive until they lose track of the bird or until the thrush finishes its journey for the night and lands somewhere to rest. Dawn may find the men anywhere from two hundred to four hundred miles from home, in completely unfamiliar surroundings, searching for the bird. To gain any useful data, they must recapture their thrush and take a small blood sample to determine how much of the doubly labeled water has been metabolized. With this information, the scientists can calculate the energetic costs of migration. Understanding the energetic demands of long-distance migration may prove crucial to determining how many rest or refueling stops the birds require while traveling between their breeding grounds and wintering grounds. Migratory birds fatten up prior to undertaking their long journeys—it's akin to filling the car up with gasoline before a trip. Wikelski wants to figure out how rapidly they burn up that fat while migrating—in essence, how many miles per gallon they get.

A migrating Swainson's thrush typically puts down before dawn, usually in the shelter of some trees or bushes. In the vast agricultural expanses of the Midwest, the only available shelter may be the shrubbery and trees growing in the backyard of a farmhouse. Such is the case on this particular morning, when Wikelski and company find themselves parked outside a house in the middle of nowhere. What follows is arguably the most difficult part of the experiment: summoning the courage to knock on the door of some stranger's house at six in the morning to request permission to set up a line of nets in the backyard for the purposes of capturing a wayward thrush. Needless to say, the ability to charm a suspicious Midwesterner at an ungodly hour of the morning is one of the prerequisites for this sort of work.

There are two great riddles surrounding bird migration: why do birds undertake such long journeys, and how they do it? The "why" riddle is usually explained in terms of opportunism: the birds are taking advantage of the incredible flush of insects produced each summer in the northern latitudes. (In the southern latitudes, a mirror-image migration occurs: some birds that breed in southern South America or southern Africa head north for the austral winter.) But given the challenges associated with flying thousands of miles each year, one cannot help but wonder whether the effort and perils of long-distance migration outweigh the benefits. Wikelski's research provides at least a partial answer. It turns out that the energetic demands of migration are not as great as once believed. An hour of flying is certainly much costlier from an energetic perspective than an hour spent resting in the forest, but only by a factor of four or five (not the twelvefold increase in metabolic demand previously assumed). As a result, the thrushes do not expend that much more energy flying than they would if they simply spent the night huddled in a woodlot in central Illinois.[2] In fact, Wikelski has calculated that actual flight constitutes less than a third of the total energy expenditure of a migrating Swainson's thrush as it makes its way from Brazil to Canada.

During its spring migration, the thrush also benefits from the prevailing winds over the eastern Caribbean and the southeastern

United States, which are mostly from the southeast. Having a tail-wind for much of its journey increases both the speed and fuel efficiency of the migrating thrush. (In contrast, during the fall the prevailing winds tend to be headwinds, which reduce the birds' speed and increase energy expenditure. Fall migrants, therefore, often must wait several days until a cold front sweeps down from the north and delivers a good tailwind.)[3]

There is little question that long-distance migration has been a very successful strategy for birds. Populations of many of North America's warblers, vireos, thrushes, and other migratory species number in the tens or even hundreds of millions. In the boreal forests of New England or Canada, they may outnumber nonmigratory songbirds by a ratio of ten to one during the breeding season. Yet long-distance migrants typically lay fewer eggs per clutch and raise fewer clutches per year than do nonmigratory birds and short-distance migrants such as robins and cardinals. For this population differential to make any sense mathematically, the long-distance migrants must, on average, outlive the residents, which means they must have a reasonably high rate of survival during migration.

Migration is also a very old strategy among birds. *Hesperornis*, a genus of flightless diving birds that roamed the seas during the Cretaceous Period some sixty-five million to eighty million years ago, was probably migratory. Its fossil remains have been found in areas that could not have sustained open water year-round, suggesting that it swam to warmer climes during the winter.[4]

Yet the migratory lifestyle, which has been a resounding success for trillions of birds for millions of years, increasingly looks like a losing proposition. Evidence suggests that populations of migratory birds in both the New and Old Worlds have plummeted in recent decades, a decline that carries the potential for significant ecological damage far beyond the loss of the birds themselves.

The evidence for this decline, however, is complex and confusing. In North America, there seems to be a near-unanimous sentiment among birdwatchers over the age of forty or fifty that the phenomenon of migration just isn't what it used to be. Veteran birdwatchers, myself included, look up at the trees and recall a time when the springtime forests rang with the songs of far more warblers, thrushes, vireos, and tanagers than one currently encounters. The trouble with such reminiscences is that they are based on hazy memories rather than hard numbers; their scientific value is negligible. (That so many birdwatchers believe there has been a decline seems noteworthy nonetheless.)

More convincing are long-term census data demonstrating a drop in migratory bird numbers across large portions of the globe. Unfortunately, such data are few and far between. Most nations or states do not have any sort of comprehensive monitoring program for birds, and the few such programs in existence are relatively recent. Within the United States, only a handful of parks and nature preserves have bird census data going back more than a decade or two, but these few, precious sites provide some important clues as to what may be happening to migratory bird populations.[5]

Consider the case of Rock Creek Park, one of Washington, D.C.'s oldest and finest parks. It remains an island of green in a region of sprawling suburbs, ubiquitous shopping malls, and expanding freeways. Since the late 1940s, a succession of dedicated birdwatchers has counted the breeding birds within an eighty-acre section of the park. Their data reveal an ominous drop in populations of songbirds that breed there but then winter in Latin America and the Caribbean. The black-and-white warbler, hooded warbler, and Kentucky warbler no longer nest in the park. Populations of ovenbirds and red-eyed vireos, once two of Rock Creek's most abundant breeding birds, have dropped by approximately 90 percent. In total, the number of migratory songbirds breeding in Rock Creek Park has dropped by 70 percent over the past half century. Conversely, the park's nonmigratory species, such as the Carolina chickadee, tufted titmouse, and downy woodpecker, have either increased or remained steady.

The situation in Rock Creek Park would be less troubling if it were an isolated case. But, in fact, most of the parks with long-term census data show similar declines in migratory songbirds. The forests of the Greenbrook Sanctuary in New Jersey, for example, harbored dozens of ovenbirds and redstarts in the 1950s; both species are now gone, and populations of other migrants in the sanctuary have dropped precipitously. Cabin John Island, in the Potomac River just outside Washington, D.C., has lost roughly half its migratory songbirds since the late 1940s.

These long-term census data indisputably show a sharp decline in populations of migratory songbirds, but we are talking about a handful of places, most of which are isolated parks surrounded by cities and suburbs. Does the situation in Rock Creek Park or Greenbrook Sanctuary reflect a broader pattern of decline among migratory songbirds, or is it purely a local issue?

Ornithologist Sidney Gauthreaux would argue that something bigger is afoot. For decades he has monitored bird migration using radar. Every spring and fall, millions of migrating songbirds cross the Gulf of Mexico, en route to their breeding or wintering grounds. So many birds pass over the gulf on certain nights that they appear as fast-moving blotches on weather radar screens. Although it is impossible to count the actual number of birds from these radar images, it is possible to tally the number of blotches and derive a crude measure of the extent of the migration. And because all the radar screens are regularly photographed and the images archived, one can look for trends. Comparing images from radar stations in Louisiana and Texas for the years 1963–67 and 1987–89, Gauthreaux came to a startling conclusion: during this period, the number of flights of migratory birds had dropped by nearly 50 percent. Moreover, in the 1960s, large numbers of songbirds were migrating almost every night when weather conditions were favorable. By the 1980s, such large flights were occurring on only 36–53 percent of the acceptable flying nights.

Gauthreaux's study would appear to provide solid evidence of a long-term decline in migratory songbird populations—what better evidence could there be than a measurement of the actual migration?—but even so, there are confounding factors that make simple interpretations risky.[6] For example, the data analyzed thus far come from only two weather stations along the gulf. Is it possible that the other stations were recording higher than normal numbers of migrants? Might the birds have shifted their migratory pathways, and, if so, could this account for the observed decline? Were the late 1980s an unusually bad time for migratory birds? No one can be certain.

Another line of evidence pointing to a decline in migratory songbirds is negative evidence. In much of the eastern United States, birdwatchers have noted a curious anomaly in the distribution of songbirds. Many migratory species do not nest in woodlots below a certain size. This threshold varies from species to species and from region to region, but as a general rule, woodlots smaller than a few hundred acres typically are "missing" a number of warblers, vireos, and thrushes that might otherwise be expected to breed there. The absence of various species from small woodlots raises a crucial question: what is it about small woodlots that makes them unsuitable for these birds? Ornithologists have been studying this question for over two decades. Their research has shown that as forests are fragmented—as they become islands of green in a paved or plowed sea—they undergo a series of ecological changes that can render them increasingly unsuitable, even dangerous, to many birds.[7]

To begin with, the developed lands surrounding many woodlots support large populations of animals that prey upon the eggs or nestlings of forest songbirds. Raccoons, blue jays, grackles, and crows all thrive in suburban environments, eating our garbage or our birdseed, and dining opportunistically on songbird eggs and nestlings. A few centuries ago, populations of these animals were probably kept in check by larger predators, such as mountain lions, wolves, hawks, and owls, but in many parts of eastern North America these top predators vanished as the forests gave way to farms and then suburbs. Add the tens of millions of dogs and cats we keep as pets (and that often become feral), and one can safely assume that songbirds in small, suburban woodlots face predation pressures of unprecedented magnitude.

This is not to say that small woodlots and backyards are devoid of birds, for they obviously are not. Rather, the mix of species is different from what one encounters in a pristine forest. Chickadees, downy woodpeckers, song sparrows, and robins all seem to thrive in close proximity to humans, whereas other birds—primarily migratory species such as warblers, vireos, and tanagers—do not. This difference may have an ecological basis. Most long-distance migrants construct open, cuplike nests, which are especially vulnerable to predators. In contrast, woodpeckers, chickadees, and titmice nest in the comparative security of tree holes.[8] Also, many long-distance migrants raise only a single brood per year before they must prepare for their fall migration. More predators mean a greater likelihood those offspring will be killed. Song sparrows, robins, and other species that do not have to gear up for a long migration can produce as many as three clutches per year, increasing the chances that at least some of their offspring will survive long enough to reach maturity and breed.

In North America, the brown-headed cowbird poses a different sort of threat to songbirds. Members of the blackbird family, cowbirds are brood parasites, laying their eggs in the nests of other species at the expense of the hosts' own eggs and offspring. Migratory songbirds are among the cowbird's favorite victims.

Prior to the arrival of European settlers, cowbirds were largely confined to the grasslands of the midcontinent, where they followed the herds of bison and other grazing mammals, eating the insects the big mammals stirred up as they walked through the grass. Because the bison were nomadic, the cowbirds could not afford to be tied down to their nests, hence the evolution of their parasitic behavior. With the clearing of the eastern forests by white settlers and the introduction of livestock, cowbirds spread throughout the eastern United States and Canada. Today they are a common sight perched jauntily on the backs of cows or chasing insects at the feet of horses from Quebec to Florida. Over the past sixty years, cowbird populations have increased further as a result of an increase in an important food resource: waste grain in southern rice fields. The advent of mechanical harvesters has simultaneously increased the amount of land under rice cultivation and the amount of waste grain left in the fields after harvest, much to the benefit of cowbirds and other blackbirds wintering in the Southeast.[9]

Thus, two factors—a range expansion and a population increase—have brought more cowbirds into contact with forest-dwelling songbirds. Many of the bird species that evolved alongside cowbirds in the Great Plains have developed defensive mechanisms to counteract their parasitic neighbors—they recognize a cowbird egg in their nest as something foreign and will either toss it out or construct a new nest and start over. But the forest songbirds, most of which happen to be migratory, fail to recognize the cowbird egg as anything out of the ordinary and consequently end up raising cowbird chicks instead of their own progeny. For years, ornithologists have attributed the inability of forest-dwelling songbirds to discriminate between their own eggs and those of the cowbirds to the fact that the forest songbirds evolved in an environment largely devoid of cowbirds; their contact with the parasitic cowbirds is relatively recent, and they have not had time to evolve effective defensive behaviors. However, there may be another, more sinister explanation for their lack of discrimination: the cowbirds apparently engage in mafia-like behavior to enforce compliance.[10] In Illinois, ornithologists Jeffrey Hoover and Scott Robinson studied the relationship between cowbirds and prothonotary warblers, a golden-yellow songbird that lives in swamp forests. They discovered that if a prothonotary warbler dares to remove a cowbird egg from its nest, the female cowbird is very likely to return to and destroy the rest of the warbler's eggs. Thus, the host ends up paying a higher price if it evicts the cowbird egg than it does if it accepts the egg and raises the cowbird's offspring as its own.

Whether the cowbirds are relaying on intimidation tactics or on the inherent naiveté of forest songbirds, they are certainly doing well. Studies have revealed staggeringly high rates of cowbird parasitism in certain parts of the country where the forest cover is now highly fragmented. In the Shawnee National Forest in southern Illinois, for example, an area riddled with clearings, roads, inholdings, and other open areas that attract cowbirds, Robinson discovered that up to three-quarters of all songbird nests were parasitized by cowbirds. Among wood thrush nests, the parasitism rate exceeded 90 percent.

Cowbird parasitism has driven a few migratory songbirds with small breeding ranges to the brink of extinction, including Kirtland's warblers in Michigan, black-capped vireos in Texas and Oklahoma, and least Bell's vireos in southern California. Only a combination of habitat restoration and continual vigilance by state and federal authorities, who trap and remove cowbirds from key breeding sites, has enabled these rare songbirds to survive. Why, then, are there any migratory songbirds left in the Shawnee National Forest or other fragmented forests where rates of nest predation and parasitism seem far in excess of reproduction and where those threats are not being reduced by human intervention?

A precise answer to this riddle has eluded ornithologists for over two decades, but it may have something to do with the spatial dynamics of songbird populations. Robinson and other researchers have proposed that areas of the country with large, continuous expanses of forest—places like the southern Appalachians, the Ozarks, the northern Great Lakes—may be net "exporters" of songbirds, producing excess progeny that disperse to other, less forested regions. Most of these dispersers will settle in poor-quality habitats (places like the Shawnee, with its fragmented forests full of cowbirds and predators), and they will fail to produce enough off-spring to replace themselves. But when they die, they will be replaced by other immigrants coming from the source areas. Thus, a steady flow of immigrants from heavily forested regions ensures that the fragmented forests will continue to have at least some songbirds.

This hypothesis carries important conservation implications, for it suggests that the continued presence of songbirds in places like the Shawnee National Forest may be due to the existence of large, intact forests hundreds of miles away. Should those forests be developed or destroyed, songbird populations might collapse throughout the Midwest. This mechanism could also explain the somewhat mysterious decline of songbirds in Rock Creek Park and other nature preserves in the Washington, D.C., area. Rock Creek Park has not changed all that much since the late 1940s. While some trees have been lost to storms, disease, and old age, on the whole the forest is as intact and stately as ever. What has changed is the land around Rock Creek. In recent decades, much of the forest cover in the greater Washington-Baltimore metropolitan area has been lost to suburban sprawl. Perhaps this regional loss of forest cover, rather than any changes specific to Rock Creek Park, was the cause of the park's dramatic loss of songbirds. The region as a whole was no longer producing enough excess progeny to stock Rock Creek Park and other isolated woodlots.[11]

In contrast to migratory songbirds in eastern North America, migratory songbirds in Europe do not seem particularly sensitive to forest fragmentation. Few if any European warblers and flycatchers (wholly unrelated to their American namesakes) shun small wood-lots. In fact, most species thrive in fragmented habitats. This difference between the two continents is, at first glance, rather puzzling, since Europe, like North America, has plenty of brood parasites and nest predators. There are important historical differences, however, between the two continents.[12] First, the effects of the Pleistocene (Ice Age) glaciations may have been more severe in Europe than in North America. Virtually all Europe's forests were erased during the Ice Age. Any European birds dependent upon large, intact blocks of forest were unlikely to survive. In parts of the American South, however, forests managed to hang on through the coldest, iciest times, surely to the advantage of some forest-dwelling birds. Second, the forests of Europe were being used and abused by humans for thousands of years before people inflicted similar damage to the forests of North America. Consequently, Old World birds have had much more time to adapt to a human-dominated landscape.

What most worries European conservationists are the threats facing "their" birds at the opposite end of the migration, on the wintering grounds. Europe's migratory birds choose a variety of winter destinations. A number of species move to the southern part of the continent, in particular the Mediterranean region. A few (some shorebirds and landbirds nesting in far eastern Europe) head as far east as India or Malaysia. But most of Europe's long-distance migrants funnel into Africa.[13] There they tend to seek out those African habitats that resemble their European breeding grounds: marsh nesters head to marshes, woodland birds to woodlands, and so forth. The region that attracts by far the greatest number of Europe's landbirds is the Sahel, the narrow band of semidesert that traverses the African continent just south of the Sahara. This is hardly a place of luxuriant tropical vegetation, of towering forests with Tarzan vines. Rather, it is a brittle landscape of thorny acacias, savannas, and scrub, a region with just enough rainfall to escape being a desert. A handful of wetlands and fertile river valleys add swaths of green to an otherwise golden-brown mantle. Yet during the northern winter, when much of Europe is under a blanket of snow, the Sahel teems with birds, drawn to the region's abundant insects and seeds. These food resources, in turn, are tied to the seasonal rains. During drought years, when seed and insect production declines, the birds suffer. Ornithologists have observed that breeding populations of many European songbirds are highly correlated with the amount of rainfall that occurred the previous winter in sub-Saharan Africa. The songbirds, they believe, perish either on their African wintering grounds or en route to Europe, perhaps because the birds are unable to store enough energy in the form of fat reserves to complete their migrations. This correlation between wintering-ground rains and breeding-season population levels is critically important because it shows that songbird populations may be controlled by events on the wintering grounds as well as on the breeding grounds.

People have been an integral part of the Sahel for tens of thousands of years. But in recent decades, a rapidly growing human population has begun to take its toll on the health of the land. Farming and livestock grazing have diminished the amount of wintering habitat available to Europe's songbirds. Even worse, the phenomenon of desertification has transformed millions of acres of arid savanna into bona fide desert, much to the detriment of the region's people and wildlife. Some scientists believe such desertification is a natural phenomenon, part of a grand cycle of rainfall and drought. Others tie it to human activities: A loss of vegetation, triggered by overgrazing, fuel-wood gathering, and other activities results in less moisture in the atmosphere (because there are fewer plants respiring). Less moisture leads to less rainfall, and less rainfall, in turn, results in droughts and the die-off of still more vegetation.

Regardless of whether it is natural or human-induced or a combination of the two, desertification has resulted in major losses of wintering habitat for many European songbirds. Its effect on Europe's songbirds naturally raises the question, could New World songbirds be declining due to changes to their wintering grounds as well? Unlike their Old World counterparts, most of whom choose to spend the winter in savannas and scrublands, many North American songbirds winter in tropical forests. Much attention has been given to the massive deforestation now under way in many of the same Latin American and Caribbean nations where U.S. and Canadian songbirds winter. Could this be harming populations of New World migrants?

Of course, not all New World migrants winter in forests. Some, like the palm warbler and indigo bunting, prefer pastures, weedy fields, and other open habitats, and they stand to benefit from deforestation. But approximately 60 species of North American birds winter in mature tropical forests. These include 29 species of warblers, 5 flycatchers, 5 vireos, and assorted thrushes, cuckoos, tanagers, orioles, and grosbeaks. Of these, a few, such as the worm-eating warbler and Swainson's warbler, occur almost exclusively in mature forests, and for them, tropical deforestation would appear to pose a serious threat. The other species, however, also inhabit second-growth forests, gardens, and coffee plantations; some even seem to prefer these habitats to undisturbed forests. Does this flexibility make them less vulnerable to the effects of tropical deforestation? The answer depends upon the nature of habitat conversion in Latin America and the Caribbean. If the land is being cleared for agriculture and then abandoned, and if younger forests are able to reclaim the cleared areas, then many of these birds are likely to persist. Such is the case in Puerto Rico, which currently has more acres of forest than it did a century or more ago. But in other parts of Latin America and the Caribbean, the mature forests are being converted to cattle pastures and sugarcane fields, persistent habitats that may be fine for palm warblers and indigo buntings but are unacceptable for almost all the others.

The situation is further complicated by the ways in which the migrants distribute themselves across the tropics. Most winter in Mexico, northern Central America, and the Greater Antilles, with proportionately fewer individuals going as far as the Lesser Antilles, southern Central America, and South America. For example, in the forests of Mexico and the Bahamas, migrants from the United States and Canada make up roughly half the winter birds. Progressively more remote destinations harbor fewer and fewer migrants. In Puerto Rico, North American migrants make up only 10 to 20 percent of the winter birds; in the Lesser Antilles, this value drops to about 1 percent. Values of 20 to 40 percent have been reported in Costa Rica and Panama, but by the time one reaches Venezuela, that figure has dropped to less than 1 percent. Of course, the lower percentages do not mean that the Lesser Antilles or South America are unimportant for migratory birds (indeed, there are some species that winter exclusively in South America), but most North American migrants pack into a relatively small region close to the U.S. mainland.

Duke University ecologist John Terborgh estimates that as many as half of all the land birds that winter south of the United States funnel into just five countries—Mexico, the Bahamas, Cuba, Haiti, and the Dominican Republic. These countries offer a combined area of roughly 836,000 square miles as compared with 6,230,000 square miles for the United States and Canada south of the tree line. "The obvious and unsettling implication of this," he writes, "is that the effects of tropical habitat destruction are amplified several times; clearing one hectare [two and a half acres] of forest in Mexico is equivalent to expanding urban sprawl by perhaps five to eight hectares [twelve to twenty acres] in the Northeast."[14] Today Haiti is almost devoid of trees, having destroyed 97 percent of its original forest cover; Cuba has converted over three-quarters of its forests to sugarcane and other crops; the Dominican Republic has lost 71 percent of its forests, and Mexico has lost half. Only the Bahamas appears to have held on to most of its forests (albeit in altered condition).[15]

Thanks to recent advances in biogeochemistry, scientists are able to show how changes in wintering habitats can affect populations of North American songbirds on their breeding grounds. For example, American redstarts nest in deciduous forests in the United States and Canada and winter in Latin America and the Caribbean. The earlier a male redstart is able to return to its breeding grounds in the spring, the more likely it is to obtain a high-quality territory, to attract a mate, and to produce offspring. Redstarts that winter in coastal mangroves and moist, lowland forests retain their body weight and condition better than do redstarts wintering in dry, scrubby habitats (the sorts of habitats that often arise following deforestation). Different winter habitats contain different types of plants, and those plants contain different carbon isotopes. The plants, in turn, are eaten by insects, and the insects are eaten by redstarts. Thus, the carbon isotopes found in redstarts when they first arrive on their nesting grounds provide a clue as to where the birds spent the winter. By analyzing blood samples taken from the birds shortly after they arrive on their breeding grounds, scientists have shown that the early arriving males are indeed the ones that wintered in the better-quality habitats and that these birds fledge more young, on average, than do restarts that wintered in poorer-quality habitats.[16]

If the rapid deforestation now under way in many parts of Latin America can be halted or even reversed, if some sort of forest is able to grow on abandoned pastures and farms, then most of the migratory songbirds should do reasonably well. But if the land does not recover—if it remains as sugarcane fields or scrub or if it is turned into apartments and shopping malls—then the potential effects on migrants could be enormous.

And what about the places in between, the pockets of forest that link the breeding grounds with the wintering grounds and provide temporary food and shelter for the migrants? These places range from the backyard oak tree in Ohio that hosts a little band of warblers for a single morning in May to birdwatching meccas like

Point Pelee, Ontario, or High Island, Texas, where migratory birds by the millions funnel through during certain seasons of the year. The honest but unacceptable answer is we just don't know. A recent study concluded that 85 percent of the annual mortality of black-throated blue warblers occurs during migration, indicating how important this phase of the life cycle is to songbird populations.[17]

That migratory birds need these rest and refueling stops is certain. Wikelski's research on migrating thrushes has shown that the birds lose a significant amount of weight each night, which they attempt to regain the following day before continuing their northward journey. But understanding the dynamics of these stopover sites is tricky. Some places appear to harbor large numbers of migrants every year. Other sites may be used only occasionally, perhaps when a sudden storm forces migrants to land. In terms of conservation planning, do we discount the value of the infrequently used sites, or do we consider them analogous to lifeboats—unappreciated until needed, then critically important? The birds themselves further complicate our efforts to understand the ecology of stopover sites by exploiting a wider variety of habitats during migration than they do on their breeding or wintering grounds. Does this flexibility mean conservationists needn't worry too much about the condition of the stopover habitat? Or are some habitat types truly better than others?

Few scientists would dispute the importance of protecting oases in the Sahara (for Old World birds heading to Africa), lush streamside vegetation in the deserts of the southwestern United States (important to many migrants in the western United States), or migration hot spots like Cape May, New Jersey. But for the vast majority of stopover sites, we can only guess how vital they really are to migrants. For example, tens of millions (maybe hundreds of millions) of migratory birds are slaughtered annually along the Mediterranean coast as they seek shelter in forests and fields after an exhausting journey across the sea.[18] Everything from hawks and doves to larks and owls is fair game, and all manner of hunting implements are employed, including guns, nets, snares, and lures. Most of this killing is for kicks, but in a number of countries there exists a thriving if discrete commercial trade in wild birds, which are served in restaurants or consumed at home.

On the European side of the Mediterranean Sea, in countries such as France, Spain, Malta, Cyprus, and Italy, bird hunting is largely an autumn activity, whereas on the African side it is equally intense in the spring and fall. All this hunting has engendered a split within the European community, with environmentalists in the northern European nations denouncing the practice as barbaric and harmful, and sportsmen in the southern nations defending it as a cherished tradition. No one is certain exactly how many birds are shot each year or, more importantly, the degree to which the slaughter is affecting overall populations. The number of birds heading south each fall, swelled by the summer's reproduction, is inevitably greater than the number returning in the spring. The question is whether the birds taken by hunters in the fall would have died anyway or whether their loss ultimately leads to diminished populations on the breeding grounds in the following spring. (One would expect spring hunting to have a greater effect on breeding populations, since it is removing birds that have already survived the autumn migration and the winter, but once again we don't know enough about the factors that ultimately limit bird populations on the breeding grounds to be certain.) Bird lovers in North America are often astounded to learn that songbirds are still considered game in parts of Europe; the hunting of such birds was banned in the United States and Canada nearly a century ago.

The strongest evidence of the importance of stopover sites to migratory birds comes not from songbirds but from various species of shorebirds, most notably the red knot.[19] Knots are robin-sized sandpipers that breed in the Arctic and winter in different spots around the world. One race in particular—*Calidris canutus rufa*—

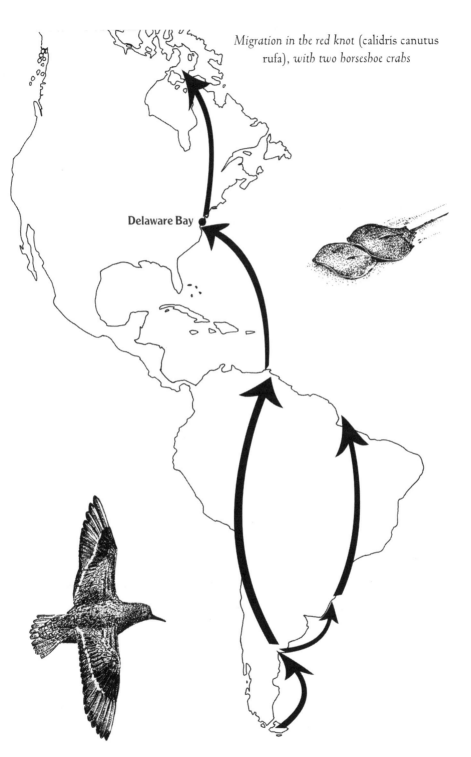

Migration in the red knot (calidris canutus rufa), with two horseshoe crabs

breeds in the northernmost parts of Canada and winters in Patagonia and Tierra del Fuego, an eighteen-thousand-mile round trip
that takes the birds from one end of the earth to the other. Refueling is critical; the knots make only a few stops along the way, but
when they do, they become voracious little eating machines, racing
back and forth, gobbling up sand crabs, amphipods, and other prey.
They must build up enough energy in the form of fat to complete
the next leg of their journey; failure to do so could be fatal. So
demanding is the journey that the knots actually shrink the size of
their intestines and other internal organs during the migratory
period in order to reduce their flying weight.[20]

Beginning in the early 1980s, biologists conducted periodic
censuses of the knots on their wintering grounds by flying along
the coastline of Chile and Argentina in a small plane and counting
the number of birds they saw. They discovered that the vast majority of knots wintered along the mudflats and beaches of Tierra del
Fuego, with a much smaller number of birds wintering as far north
as Peninsula Valdes in Patagonia.[21] Because the biologists were
able to fly over the entire winter range of the birds, they were able
to take a census of virtually the entire population. In 1985, the winter tally exceeded 67,000 individuals. The count in 2000 was down
to 51,255 birds. The 2002 census found only 29,271 knots, a further decline of over 40 percent. By 2006, the population numbered
only 17,211 individuals.

Why the sudden drop in numbers? Biologists found no evidence of any dramatic changes on either the breeding or wintering
grounds. Hunting was not a major issue; some knots are undoubtedly shot while migrating, but the birds are not hunted in great
numbers. Nor were there any indications of a sudden disease outbreak. Weather conditions during the migratory period seemed
more or less normal. Attention then turned to the stopover sites,
and one in particular: Delaware Bay.

Every year toward the end of May, virtually the entire population of *Calidris canutus rufa* funnels into Delaware Bay to rest and eat. It's the last refueling stop for the birds before they undertake the final leg of their migration, a fifteen-hundred-mile sprint to their nesting grounds in the Canadian tundra. The knots arrive in Delaware Bay exhausted and depleted, each weighing about four and a half ounces. During the span of a couple weeks, they must put on an additional one and two-thirds ounces of fat just to fly to Canada. Moreover, because the tundra may still be frozen when they arrive on their breeding grounds, preventing them from finding food, the knots will need some additional fat to sustain themselves for a few days until conditions improve. All told, then, they may have to increase their body weight by over 50 percent while in Delaware Bay.

Their ability to go on a sustained eating binge while at the bay is due entirely to another creature: the horseshoe crab. These bizarre animals are among the oldest on earth. Horseshoe crab fossils over 200 million years old are almost identical to contemporary animals; they are the closest living relatives of the trilobites. Each spring, millions of horseshoe crabs migrate from the waters of the continental shelf to the shallower waters of Delaware Bay to spawn. Males literally attach themselves to the females (an arrangement known as amplexus), and the pairs then make their way to the sandy intertidal zone. Each female digs a shallow pit in the sand, where she lays tens of thousands of tiny, blue-green eggs, usually in clusters of a few thousand.[22] Wave action and the digging of nearby horseshoe crabs disturb many of these egg clusters, bringing them closer to the surface. It is this ephemeral resource—billions of eggs, some lying just beneath the surface of the sand, others churned up by the waves and deposited in long wrack lines on the shore—that attracts hundreds of thousands of hungry gulls and sandpipers, including red knots. The result is an avian feeding frenzy that ranks as one of the greatest wildlife spectacles of North America.

By trapping, marking, and weighing the knots, biologists have been able to piece together some alarming trends in Delaware Bay.

First, the proportion of birds that had gained the requisite weight to successfully complete their northward journey dropped markedly in 2002. Many of the birds departing Delaware Bay at the end of May that year were underweight, and relatively few of them reappeared the following spring, indicating high mortality over the course of the year. Second, the number of young knots appearing in Delaware Bay in subsequent years also dropped, an indication of low breeding success. Finally, a handful of adult knots were captured, killed, and dissected in the spring by the scientists, who were alarmed to discover that the birds were not regrowing their intestines and other internal organs as quickly as expected. (The knots need to have their internal organs back in shape by the time they arrive on the breeding grounds). All these clues suggested that Delaware Bay was no longer the refueling stop it once had been.

And indeed it wasn't. During the 1990s, commercial fishing for horseshoe crabs increased dramatically. People were capturing the crabs and selling them as bait for the more profitable eel and whelk fisheries. For some reason, the odor of egg-laden female horseshoe crabs (or pieces thereof) is deeply alluring to eels; either male or female horseshoe crabs work well for the less picky whelks.[23] The annual harvest of horseshoe crabs rose from approximately 800,000 pounds in 1993 to nearly 6.4 million pounds in 1998. Over the same time period, however, the density of crabs (measured in terms of the number captured per net tow) fell sharply. The inescapable conclusion is that fishers were depleting Delaware Bay's horseshoe crab population, causing a reduction in the number of horseshoe crab eggs upon which the shorebirds fed.[24]

Faced with evidence that overfishing of the horseshoe crabs was harming the knots, the states lining Delaware Bay initially did what one would expect them to do: they ducked the issue, calling for more studies before committing to any steps that might anger the fishing industry. As the studies all pointed to the same conclusion, however, and as the birds failed to rebound, state and federal authorities began to take action. They reduced the allowable harvest of horseshoe crabs, and they prohibited any harvest at all during the period when the shorebirds are visiting. As a consequence, the overall harvest of horseshoe crabs dropped by 62 percent between 1998 and 2003.[25] The states of New Jersey and Delaware went even further: in 2006, they declared a two-year moratorium on the harvest of all horseshoe crabs.[26]

It remains to be seen whether the reduced harvest and closed season will suffice to bring about a revival of the horseshoe crabs. Even if these measures work, recovery of the horseshoe crab population may take years, given that female crabs do not reach sexual maturity until they are nearly a decade old.

There is, of course, every reason to hope these steps will succeed, not only for the sake of the knots, but also for the hundreds of thousands of other birds that partake of the bay's brief bounty of crab eggs. Hoards of ruddy turnstones, semipalmated sandpipers, sanderlings, short-billed dowitchers, and other migratory shorebirds also pour into Delaware Bay during this same brief period in the spring. All are headed to nesting grounds in the Arctic, and all need to fatten up for their northward journey.

For well over a decade ornithologists have been sounding alarms over seemingly widespread declines among many of North America's shorebirds, not just the red knot. A study published in 2001 identified nineteen species showing statistically significant or persistent population declines and only one species showing a significant increase.[27] "Declines in shorebird populations," the authors concluded, "appear to be even more extensive and severe than previously thought, emphasizing the urgent need for conservation measures for this group of birds in the Western Hemisphere."[28] Of course, the troubles in Delaware Bay cannot explain the declines of all these species, but the loss or degradation of key stopover sites, which are scattered across North and South America, is surely an important factor for some.

I first witnessed the shorebird congregation in Delaware Bay in 1987, having been drawn to the site by an article in *Natural History* magazine with the alluring title "Sex and Gluttony on Delaware Bay."[29] What I encountered was extraordinary. Scattered all along the shore were the shells of dead and dying horseshoe crabs, dark lumps the size of dinner plates. A few remained in their copulatory embrace, apparently driven to complete their mission even as the sun baked them to death. Thousands of aptly named laughing gulls created a background chorus of guffaws, while clouds of red knots, semipalmated sandpipers, ruddy turnstones, and other shorebirds alit upon the beach, frantically probing the sand and mud for crab eggs, until a passing hawk or human flushed the birds, causing them to wheel and turn in unison until they once again came down to continue their nervous feasting. The sex and gluttony were both there, along with great beauty. I felt as though I had stepped into the shoes of John James Audubon, back into an era of wilderness and abundant wildlife, yet here I was in New Jersey, a half hour's drive from Atlantic City.

Nearly twenty years later, I returned with some students, eager to experience the phenomenon again. The students, none of whom had been to Delaware Bay before, were dazzled, yet I was disappointed. All the species were there—the knots, the turnstones, and other sandpipers—and they were delightful to watch, but the numbers seemed just a fraction of what I remembered from my earlier visit.

THOSE OF US LIVING in the northern latitudes naturally think of bird migration (and, consequently, bird conservation) as a north–south affair. But the diversity of migratory pathways among birds nearly rivals the diversity of birds. Many African larks and pipits, for example, never leave the continent but wander widely within it, tracking the ephemeral rains that bring new plant growth and a flush of insects. There are flycatchers and swallows in Chile and Argentina that escape the austral winter by flying north to the Amazon. And in recent decades scientists have learned that many tropical birds travel up and down mountain ranges in response to seasonal fluctuations in the availability of fruit and nectar. Although their journeys may be shorter than many of the north–south migrants, their needs are the same: adequate breeding habitat, a secure winter home, and enough stopover sites in between. Unfortunately, we know little about the migratory pathways of these other migrants and are therefore even less equipped to protect them than is the case for "traditional" north–south migrants.

One of the few tropical migrants whose travels have been studied in some detail is the three-wattled bellbird. The male bellbird is a hefty, pigeon-sized bird with an immaculate white head and a bright chestnut-colored body. It also sports three black, wormlike wattles that hang from the base of its bill and serve mainly to annoy ornithologists, who can think of no earthly reason why the male bellbird should have them. The female bellbird lacks the wattles and is more modestly colored in shades of olive and yellow. Names can be misleading, none more so than in the case of this species, whose call sounds nothing at all like a bell. Rather, the male emits a painfully loud clanging sound, reminiscent of a blacksmith's hammer hitting a forge, that can easily be heard for a half mile or more.

Three-wattled bellbirds nest in montane cloud forests from Honduras to western Panama. Ornithologists have long known that bellbirds engage in some sort of altitudinal migration inasmuch as they disappear from the mountains during certain times of the year and suddenly appear in the lowlands, but the full extent of their migration did not become apparent until a team of scientists under the direction of George Powell began placing radio transmitters on bellbirds in Costa Rica and tracking their movements. Powell's team discovered that Costa Rica's bellbirds undertake one of the most complex migrations ever recorded for a tropical bird.[30]

From March through June, the bellbirds bred in middle-elevation forests along the Atlantic slope of the Cordillera de Tilarán in north-central Costa Rica. In June and July, after nesting, they crossed the continental divide and took up residence in middle-elevation forests along the Pacific slope of the cordillera. Then, in September and October, the birds headed to low-elevation forests in southeastern Nicaragua and northeastern Costa Rica. A couple months later, in November and December, they migrated to forests along the southwestern coast of Costa Rica. By March, they were headed back to their breeding grounds in midelevation forests on the Atlantic side of the Cordillera de Tilarán. Thus, in the course of a year, the bellbirds wandered from midelevation to low-elevation forests and back to the midelevation forests, and they repeatedly crossed the continental divide. The likely driving force behind this peculiar odyssey is fruit. Bellbirds live almost entirely on fruit, which is an ephemeral resource in the tropics. The fruiting trees favored by bellbirds come into season at different times of the year at different elevations and latitudes. The bellbirds follow the fruit, in effect adopting a migratory lifestyle that corresponds with the fruiting times of the trees.

The bad news from a conservation perspective is that some of the areas where the bellbirds travel—for example, the middle-elevation zone along the Pacific slope of the Cordillera de Tilarán and the lowland region of southwestern Costa Rica—are largely deforested due to agriculture, leaving the bellbirds with very little habitat. Costa Rica's exemplary system of national parks and nature reserves does not include much land in these areas, an omission that could condemn the bellbirds to eventual extinction.

The aesthetic loss would be enormous, as anyone fortunate enough to see a three-wattled bellbird can attest. But the consequences extend well beyond aesthetics. Fruit-eating birds play an essential role in the dispersal of fruit-bearing trees. By consuming fruit and subsequently defecating the seeds, the birds ensure that the seeds are carried to new locations away from the parent tree, to places where they have a better chance of growing. Male bellbirds are especially useful in this regard because, after dining at a fruiting tree, they typically return to a favored perch in the canopy from which they spend hours and hours calling and defecating. Because bellbirds favor perches along the edges of openings or gaps in the canopy, the seeds they defecate are more likely to wind up in an open, sunny area. These are precisely the conditions that the seeds of some tree species need to germinate. The bellbirds, in other words, are "directed dispersers," unintentionally but successfully placing seeds in the right places to grow.[31] Thus, the disappearance of the bellbirds could conceivably interfere with the reproduction of various tree species. The demise of these fruit-bearing trees, in turn, would be harmful to the many other animals besides bellbirds that consume fruit.

The same Costa Rican cloud forests harboring bellbirds also contain another remarkable animal facing a similar plight: the resplendent quetzal, that legendary bird whose iridescent colors and exquisite, long plumes have made it a deity to the Mayans and a holy grail to today's birdwatchers. It too engages in a complex altitudinal migration in search of fruiting trees (principally avocados), and many of the forests visited by migrating quetzals are unprotected and vulnerable to destruction. The loss of these forests could lead to the extirpation of the birds. This would be an economic disaster for communities like Monteverde, Costa Rica, which derives millions of dollars annually from visiting birdwatchers drawn to the area by the prospect of seeing wild quetzals.[32]

THE GATHERING OF SHOREBIRDS in Delaware Bay, the seasonal wanderings of bellbirds and quetzals in Central America, and the northward migration of songbirds in the Midwest are all examples of migrations timed to correspond to other natural phenomena over which the birds themselves have no control. Shorebirds, for example, must arrive in Delaware Bay shortly after the horseshoe crabs have spawned. If the birds arrive too early or too late, the key food resource (crab eggs) may not be there. Bellbirds and quetzals track the fruiting patterns of different species of trees at different elevations. They cannot directly control when or where the trees come into fruit (which, presumably, is tied to environmental factors such as rainfall and solar radiation), but they do need to be there at the right time nonetheless.[33] In the case of songbirds passing through the Midwest en route to their breeding grounds in Minnesota or Manitoba, their journey is tied to the leafing out of the trees, especially the oaks. When the new leaves emerge, they are low in tannins and other chemicals that protect them against insects. Consequently, a variety of moths lay their eggs on the new leaves, resulting in a bounty of caterpillars by the time the warblers pass through. For the warblers, the abundant caterpillars are just what they need to rebuild their energy reserves in order to continue their northward migration.

An appreciative naturalist cannot help but marvel at the exquisite timing that underlies these migratory phenomena. That natural selection, operating through endless iterations of trial and error, can link together the life cycles of creatures as different as knots and horseshoe crabs, bellbirds and avocado trees, and blackburnian warblers and caterpillars is nothing short of amazing. But it is also a cause for concern. Given that humans are recklessly engaged in the process of changing the earth's climate through the production of greenhouse gases, how will these complex migratory patterns fare? Could the birds become decoupled from their food resources, and, if so, how might that affect them and their ecosystems?

Paul Strode, an ecologist at the University of Illinois, has been pondering this very question with respect to midwestern songbirds. A couple years ago, he invited me to hop aboard a cherry picker and join him for a bird's-eye view of the woodlots of central

Illinois. Once the machine had lifted us into the canopy, Strode began counting caterpillars on the branches of various trees. His goal was to determine when the caterpillars appeared, how important they were to migrating warblers, and which types of trees harbored the most caterpillars. Around some branches, Strode had placed mesh netting to keep the birds away. By comparing caterpillar numbers on these protected branches with caterpillar numbers on branches of the same tree that were accessible to the warblers, he could determine whether the birds truly made a dent in caterpillar populations. Strode also combed through meteorological records and the diaries of birdwatchers to determine the timing of both leaf emergence and the arrival of the warblers.

His preliminary findings justify at least a moderate degree of alarm.[34] Over the past century, springlike conditions that favor the emergence of leaves (and, consequently, the emergence of caterpillars) appear to be arriving later and later in southern Illinois. In northern Minnesota, in contrast, springlike conditions are arriving earlier, presumably resulting in the earlier appearance of caterpillars on the warblers' breeding grounds. The migrating warblers, therefore, are caught in a bind. They need to get to their breeding grounds sooner to catch the peak of the caterpillar outbreak. If they arrive too late, they may find that the caterpillars have grown too large to capture easily or, worse yet, they may have already metamorphosed into moths. Yet along part of the migratory route, spring is arriving later, meaning there may not be enough caterpillars at the Midwest stopover sites to allow the birds to gain enough energy to complete their migration successfully. Strode has calculated that, because of climate change, warblers now have up to twenty fewer days to travel between southern Illinois and northern Minnesota. That's certainly bad news for the birds and potentially bad news for Minnesota's forests. Studies elsewhere have shown that insectivorous birds help to control populations of defoliating caterpillars. If the warblers are late in reaching their breeding grounds, thus giving the caterpillars more time to grow and eat, the trees may suffer as a result.

In the Netherlands, global warming already appears to have taken a toll on populations of the pied flycatcher, a small black-and-white bird that breeds in Europe and winters in Africa.35 Pied flycatchers depend upon an abundant supply of caterpillars to feed their nestlings. In apparent response to rising temperatures in the Netherlands, the caterpillars have been emerging earlier and earlier. The flycatchers, in turn, have been laying their eggs earlier. However, they have not been able to push ahead the time of their arrival in the Netherlands, perhaps because the genetic cue to leave their winter quarters and head north is based on some fixed parameter, such as seasonal changes in day length. The birds can shorten the time period between arrival on the breeding grounds and egg laying by only so much, since they must recover from their long migration and get into shape for breeding. Consequently, the peak of the caterpillar population no longer corresponds to the time when the flycatchers are provisioning nestlings. In those parts of the Netherlands with the greatest divergence between caterpillar emergence and flycatcher arrival, populations of the flycatchers have declined by about 90 percent over the past two decades.

Elsewhere, we know very little about the influences of climate on such phenomena as the spawning of horseshoe crabs or the fruiting of tropical trees, but one cannot help but wonder whether these migratory linkages are also at risk due to global warming—and how many other, as-yet-undiscovered linkages may be severed to the detriment of migratory species.

GIVEN THE PROPENSITY of migratory birds to cross all sorts of national and international borders, the political landscape becomes nearly as important to these birds' survival as the physical landscape. The challenges associated with bringing different agencies, institutions, states, provinces, and nations to the table in order to hammer out a comprehensive strategy for protecting migratory birds are undeniably daunting, but it can be done. For example, Canadian and U.S. wildlife agencies have worked together for decades to coordinate management activities and bag limits for migratory waterfowl. Growing concern over the plight of migratory songbirds led to the formation of a new conservation coalition in North America in 1990 called Partners in Flight.[36] Partners in Flight bills itself as the world's largest avian conservation program, a statement of some hubris that nonetheless contains a kernel of truth. The singular accomplishment of Partners in Flight has been to bring virtually everybody to the table: federal agencies, state agencies, not-for-profit conservation organizations, corporate landowners, academic scientists, and birdwatchers. More precisely, it has spawned a multitude of committees and working groups that meet endlessly but that also produce valuable conservation strategies and research agendas for different parts of North America. In any endeavor of this sort, the process is as important as the product: by bringing together so many players, Partners in Flight has raised the profile of migratory birds within state and federal agencies that had previously given them short shrift. Among the European nations, a Directive on the Conservation of Wild Birds was signed in 1979 covering the "protection, management, and control" of all wild bird species native to Europe, and a number of the signatories have protected important habitats or enacted new hunting regulations in response to perceived threats to migratory birds.[37]

There is even an international treaty that aims to protect not only birds, but all types of migratory animals. The Convention on the Conservation of Migratory Species of Wild Animals, also know as the Bonn Convention, was signed in 1979. Its 102 signatories (unfortunately not including the United States, Canada, or Mexico) have agreed to identify and protect migratory species of all types in a coordinated fashion. Meetings take place, vulnerable species are

Sandhill cranes

identified, and threats are discussed, but the Bonn Convention nonetheless has been a disappointment. Only a handful of conservation agreements covering small numbers of species have been finalized. Moreover, since the convention itself does not specify what conservation measures must be taken by the signatories, each agreement is only as strong as the participating nations choose to make it.[38] The Bonn Convention is a start, certainly, but only a start.

To no one's surprise, the developing nations in Asia, Africa, and Latin America have had a harder time coming up with coordinated plans to protect their migratory birds (many of which are also our birds, if "our" is defined as North America or Europe). Chronically underfunded, facing immense and immediate challenges of poverty and political instability, these countries simply cannot afford to give birds the same attention they receive in the northern nations. That said, a surprising number of cooperative research, monitoring, and education programs focused on migratory birds have sprung up between rich and poor countries. If not enough to justify outright optimism, these joint ventures at least are sufficient to ward off entrenched pessimism.

Besides, whatever pessimism I might feel in the face of all the depressing data simply melts away each spring as the land and the skies reawaken. A few years ago, in the predawn blackness of a cold March morning, a friend and I stepped into a blind at the edge of the Platte River in central Nebraska. Although we could see nothing in the darkness, occasional throaty growls and trumpetlike bugles told us we were in the right place. All around us, we knew, were cranes, lots and lots of sandhill cranes. Every spring, a half million or more sandhill cranes—80 percent of the world's population—funnel into this small stretch of the Platte. Drawn from throughout their wintering grounds in Mexico and the southwestern United States, the cranes pause here for a few weeks before heading on to their breeding grounds in Canada, Alaska, and even Siberia. While in Nebraska, they spend their days foraging in farm fields and nearby wetlands. At dusk, they gather together in immense flocks and settle down to roost in a few shallow stretches of the Platte.

As dawn broke, we realized that the stretch of the river in front of us was covered in cranes, tens of thousands of sandhill cranes packed together like sardines, stretching downstream for at least a mile. The majority of the birds were still asleep, their bills neatly tucked into the feathers of their backs, but a growing number were waking up, stretching their necks and wings, ruffling their feathers, and occasionally trumpeting a greeting call to a nearby neighbor.

By 7:00 a.m., all the cranes were awake and active. Some were squabbling with each other, jabbing at their neighbors in retaliation for real or perceived incursions into their personal space. Others were courting, bowing their long necks while simultaneously raising their wings above their backs, then catapulting into the air and landing gently. By the minute, more and more birds stretched their broad wings, ran a few steps in the shallow water, and took flight, calling loudly as they headed off to the farm fields in search of food. In every direction, the sky was filled with lines of cranes. It was a rush hour like no other in the world.

By 7:45, approximately half of the birds had left. Suddenly and without warning, all the remaining birds on the river took to the air, a mile-long blanket of cranes rising from the water and filling the sky, all bugling loudly as they disappeared over the horizon. The din was stupendous, like the roar of a packed stadium when the hometown team scores the winning touchdown. And within minutes, the exodus was over, and the river lay empty and quiet. I could hardly believe that the same scene would replay itself tomorrow and the next day and the day after that, as it has every March for thousands of years.

CHAPTER 2

A Mountain of Butterflies and a Cloud of Grasshoppers

Every autumn, Cape May, New Jersey, becomes a mecca for birdwatchers. Migrating birds moving down the coast reach the tip of the Cape May Peninsula and stall, apparently reluctant to cross the waters of Delaware Bay. They mill about Cape May's forests, fields, and marshes for a few hours or a few days, resting and feeding before braving the Delaware crossing and continuing their southward journey. Under the right winds and weather conditions, Cape May can host hundreds of thousands, even millions, of migratory birds.

Birdwatchers gather in flocks to enjoy the spectacle, tallying the hawks, warblers, vireos, flycatchers, tanagers, swallows, and other species. I have been going there for years, but it wasn't until just a few years ago that I began to notice that birds weren't the only creatures on the move at Cape May. There were also dragon-flies and butterflies. At certain times in September, thousands of orange and black monarch butterflies floated across the sky or drank nectar from goldenrods and asters in the meadows; on other occasions, swarms of dragonflies milled around the sand dunes or moved en masse up the coast. On the afternoon of September 11, 1992, for example, one biologist estimated over four hundred thou-sand dragonflies passing by Cape May Point during a seventy-five-minute observation period. The vast majority were green darn-ers—a beautiful green and blue species—along with smaller num-bers of black saddlebags (named for the dark patches at the base of their hind wings) and twelve-spotted skimmers (named for the spots on their wings).[1]

Insect migration is hardly unique to Cape May. Around the world, numerous species of butterflies, moths, bees, flies, and other insects engage in surprisingly long journeys. Yet despite being a widespread phenomenon, insect migration remains a poorly stud-ied one, at least relative to the better-known travels of birds and mammals. What little we know suggests that the migrations of insects, like those of most birds and mammals, are driven by oppor-tunism, by the chance to take advantage of food or other resources that are not available year-round in a particular place. Moreover, there are striking similarities in the patterns and mechanisms of insect migrations compared to those of other animals. Regrettably, this includes the increased vulnerability to human disturbance that goes with a migratory lifestyle. Even the most abundant insects can be harmed by the loss of breeding, wintering, or stopover habitats.

Among migratory insects, dragonflies may be the most myste-rious. At least fifty species of dragonflies around the world are thought to be migratory. Swarms of migrating dragonflies have been sighted in North America, in Europe, over the pampas of Argentina and Uruguay (where the naturalist and writer William Henry Hudson described "dragonfly storms"), and in sub-Saharan Africa. A single swarm of migrating dragonflies observed in Argentina in December 1991 was estimated to contain four to six *billion* individuals. The density of dragonflies within such swarms easily approaches that of locusts.[2] In North America, significant migrations occur along the Atlantic and Gulf coasts, the Great Lakes, and the ridgelines of the Appalachians and Catskills.

Migrating birds can be tracked with radio transmitters, or they can be marked by placing colored bands on their legs. Either technique will shed light on their comings and goings. Dragonflies, by contrast, are much harder to tag or track. And even if one could paint an identification number or stick a label on a dragonfly's wings, the odds of resighting that individual a day or two later, perhaps in the company of tens of thousands of other dragonflies, are infinitesimally small. Thus, despite a century or more of documenting the phenomenon, scientists have virtually no clue where the dragonflies are headed or even why they are migrating.

In North America, at least, dragonfly migrations seem to follow the same patterns observed in migratory birds: the largest autumn flights take place after the passage of a cold front; the spring flights are greatest when a warm front ushers in southwesterly winds; and the migrants tend to concentrate along topographic "leading lines," such as coastlines or mountain ranges. Moreover, like migrating birds, dragonflies store substantial amounts of fat prior to travel.

There is, however, a fundamental difference between bird migration and dragonfly migration, at least in North America: the dragonflies heading south in the fall never make it back to their breeding grounds. They do not last through the winter. We know this because the ones heading north in the spring do not show much wear and tear on their wings; they appear to be individuals that have only recently metamorphosed from larvae to adults. (A dragonfly that had successfully migrated south and spent the winter somewhere would show a number of nicks and tears on its wings by the following spring.) Thus, it appears as though dragonflies are generational migrants. Individuals head south in the fall, laying eggs en route or upon reaching some final destination. These adults die sometime during the autumn or winter. Meanwhile, their eggs hatch, and the resulting larvae metamorphose into adults, who then head north in the spring. The dragonflies migrating north in the spring thus are probably the children—or perhaps even the grandchildren—of the individuals that headed south the previous fall. How all of this is coded in the genes is a complete mystery.

What little we know about dragonfly migrations is based almost entirely on coarse observations of the phenomenon itself: dragonfly watchers record the time of year and weather conditions when large swarms occur, they note the direction the swarms are headed, and they estimate the numbers or densities of dragonflies on the move. What they have not been able to do—until very recently—is follow the actual migration of an individual dragonfly, if only for a day or two.

In the fall of 2004, I took Princeton's Martin Wikelski to Cape May to see the bird and dragonfly migration. While we were watching the green darners, swamp darners, and black saddlebags zipping across the meadows, I posed a simple question to him: is there any way to track one of these insects on its southward migration? Wikelski expressed confidence that his friend Jim Cochran, a brilliant designer of electronic tracking devices, could devise a radio transmitter small enough to fit on a dragonfly.

It was a remarkably audacious statement. An adult green darner weighs no more than a gram and a half, roughly equal to one-twentieth of an ounce. Although dragonflies are capable of flying while carrying additional weight—they sometimes carry prey items as large or larger than themselves—there have to be limits to how heavy a transmitter they can carry and still migrate successfully. If one aimed to keep the weight of the transmitter to no more than 20 percent of the dragonfly's body weight (equivalent to a 150-pound hiker carrying a 30-pound pack), it would have to weigh no more than one one-hundredth of an ounce. The simplest way to reduce the weight of the transmitter is to reduce the size of the battery, but the smaller the battery, the less powerful it is, reducing the lifespan and transmission range of the transmitter. Building the right device involves numerous trade-offs, requiring a combination of engineering skills and imagination. Nonetheless, the following summer a small package—a very small package— arrived in Wikelski's office, and inside it were a half-dozen tiny transmitters, the product of a master craftsman.

That fall, Wikelski and two colleagues from Rutgers, Michael May and David Moskowitz, captured a dozen green darners in Princeton and Cape May. With a few drops of eyelash adhesive and superglue, they attached the transmitters to the insects' bellies. Much to the scientists' relief, the tagged dragonflies seemed to have no trouble flying, and they quickly resumed their normal routine. Over the next few weeks, the scientists tracked the dragonflies on foot, by car, and via a small Cessna airplane equipped with external receiver antennas. Each dragonfly was followed, on average, for six days. The resulting data, although fragmentary, provide the first glimpse into the life of a migrating insect.3

Wikelski and colleagues discovered that migrating green darners have travel days and rest days. During the latter, the insects presumably feed and build up energy for the flights to come. The longest one-day journey they recorded was over eighty-five miles, when a particularly intrepid darner flew from the north shore of Delaware Bay to the coast of Maryland. Including the rest days, the dragonflies averaged a little over seven miles per day during their time in New Jersey. That may not seem like much, but over the course of a two-month migration period, it would allow them to move over four hundred miles south.4 The dragonflies did not travel during blustery days, when they would have risked being harmed or blown in the wrong direction. They increased their likelihood of traveling in the right direction (i.e., southward) by appearing to obey a simple rule: migrate only after two nights of successively colder temperatures. Falling nighttime temperatures often signal the arrival of a cold front, which produces the northwest winds conducive to a southward migration. Dragonflies that reached the tip of the Cape May Peninsula behaved much like migrating songbirds: they stopped, reversed course, and headed north until they found a narrower stretch of the bay that they could cross.

Given the limited lifespan of the transmitters, Wikelski and colleagues could not determine how far south the insects ultimately went or how long they lived. Thus, their study is no more than a peek into the world of a migrating dragonfly, but it is a peek that many dragonfly watchers never dreamed would be possible. With smaller batteries and more powerful receivers, it may soon be possible to track a single dragonfly over the entirety of its migration. If this tracking can be coupled with ground observations, scientists should be able to discover where and when the dragonflies are reproducing, information that will go a long way to solving the riddle of why they migrate in the first place.

ALTHOUGH WE HAVE BARELY BEGUN to piece together the travels of dragonflies, the migration of another insect, the monarch butterfly, is now reasonably well understood. Yet just thirty-five years ago, it too was a great mystery. This much we knew back then: Large numbers of monarchs spent the winter in pine and eucalyptus trees along the California coast, from the Bay Area south to San Diego. Most of these wintering sites contained fewer than a thousand butterflies, but in a few places tens of thousands of monarchs crowded together. However impressive these aggregations might have seemed, they were insufficient to account for the hundreds of millions of monarchs that fill the meadows of eastern North America every summer.

There was no shortage of theories. Some naturalists thought the monarchs overwintered along the Gulf Coast in Mississippi or Louisiana. Others guessed the butterflies were spending the winter in Florida. (It's a measure of how undeveloped Florida was just a few decades ago that it was possible to imagine millions of monarchs living there undetected. In fact, Florida does harbor nonmigratory populations of monarchs, but those populations do not journey north in the spring.)[5] A few nineteenth-century naturalists even argued that the eastern monarchs had no wintering grounds but engaged in a pointless, lemminglike march to the south. According to this theory, the reappearance of monarchs every year in the northeastern United States was the result of a few hardy adults or pupae that managed to survive the northern winter, perhaps by hibernating under the bark of trees or under logs.

Starting in the 1930s, Canadian entomologist Fred Urquhart tried to solve the riddle of the monarch's winter home.[6] He realized that any serious attempt to find the wintering grounds would require him to track the movements of individual monarchs. In other words, he needed to know where individual butterflies were headed. Radio telemetry was out of the question; the technology simply did not exist (and still doesn't) for something as light as a butterfly. Instead, Urquhart took his cue from bird banding. For over a century, ornithologists had been putting metal rings on the legs of migratory birds. When these ringed birds were either recaptured or found dead at a different location, the ornithologists could reconstruct the birds' migratory route. The challenge was to devise something similar for monarchs.

Urquhart experimented with a variety of techniques before hitting upon a simple but ingenious solution: a tiny white adhesive tag, smaller than a fingernail, that could be folded over the leading edge of the forewing. On the tag was printed a unique identification number and the notation "return to Museum, Toronto, Canada" (or, in later years, other institutions).7 With this innovation, a small army of volunteers could now head into the fields in the autumn to tag monarchs. If enough butterflies were marked and enough people were on the lookout for them as they headed south, enough information might eventually be gathered to reveal the monarch's winter home.

Urquhart plotted the recoveries on a large topographic map in his office, using black thread to connect the point of initial capture with the point of recapture. After a number of years, the result was not a spider's web of crisscrossing lines, but rather a series of lines that tended to point in the same direction, toward the southwest. Most of the monarchs were headed toward the Gulf Coast and then continuing in a westerly direction into Texas. Urquhart and his wife spent the winter of 1969–70 in southern Texas, hoping the wintering grounds might lie somewhere within the Lone Star State. They eventually concluded that the butterflies were heading farther south, into Mexico and perhaps Central America.

The couple started searching the Mexican countryside but soon realized that their chances of finding the butterflies on their own in such a vast country were next to nil. They therefore changed tactics, writing numerous articles in English and Spanish for popular magazines and newspapers in Mexico. The more people who knew of their efforts to locate the monarchs, they reasoned, the greater the likelihood that someone somewhere would have a clue. Their publicity campaign inspired Kenneth Brugger, an engineer working in Mexico City, to take up the hunt. Traveling across the Mexican countryside in a mobile home, Brugger targeted those areas where tagged monarchs had been sighted as well as places where other visitors had reported high concentrations of butterflies. By the end of 1974, Brugger had narrowed his search to the vicinity of the town of Angangueo, located high in the mountains of Michoacán, approximately seventy-five miles west of Mexico City. On January 9, 1975, Brugger telephoned Urquhart with the electrifying news, "We have located the colony! We have found them—millions of monarchs—in evergreens beside a mountain clearing."[8] Urquhart himself would not see the monarchs' winter home until a year later, when Brugger took him to the site. There, Urquhart scrambled down the slopes of a dormant volcano and gazed upon a stand of ancient fir trees:

> The dense growth of spruce trees was festooned with giant clusters of monarchs that resembled so many dead leaves, for at the time not one butterfly was stirring. They blocked out the light, except for a small gap that permitted the rays of the sun to penetrate the forest floor, which looked like a gigantic Persian carpet because it, too, was covered with orange monarch butterflies. The stillness of the air, the damp, somber darkness of the forest, the blue-gray beams of light like ethereal pathways to the blue of the heavens above all gave the impression of a cathedral where one should converse in whispered tones for fear of breaking the enchantment. As we gazed in silence at the scene before us, the sun, which had been hidden behind a gray cloud, emerged and beamed a warm ray of golden light upon one of the great clusters of dormant butterflies. As if on cue from a director, the tree was transformed into a blaze of color as the butterflies spread their bright orange wings to the warm sunlight.[9]

Subsequent searches by Urquhart, Lincoln Brower, and others would uncover approximately ten other sites harboring wintering monarchs in that same small region of Mexico. The butterflies apparently funnel through Texas until they hit the mountains of the Sierra Madre Oriental. They then follow the mountains to the southeast, eventually crossing them to reach the Transvolcanic Belt, a string of volcanic mountains at the southern end of Mexico's

Central Plateau. Here, the butterflies spend the winter in a few patches of high-elevation fir forests. Within these sites, the butterflies cluster together in colonies ranging in size from a quarter acre to fourteen acres, at densities of over twenty million butterflies per acre.[10]

Urquhart had expected the monarchs to be in the warm, sunny lowlands of Mexico or Central America, sipping the nectar of tropical flowers. Instead, they were quite literally shivering in fir forests at an altitude of ten thousand to twelve thousand feet, in places where nighttime temperatures come close to freezing. From time to time, these places are even hit with snowstorms, raising the obvious question, why would the butterflies spend the winter in such a seemingly hostile environment? Why not do what Urquhart and others expected them to do—stay in the lowlands? The answer may lie in the challenges associated with the monarch's unique migration.[11] The butterflies accumulate fat reserves during their fall migration, especially when they reach Texas and northern Mexico.[12] They then use this fat to get through the winter in Michoacán. Being cold-blooded creatures, their metabolic rate, and hence the rate at which they deplete their fat reserves, depend upon the ambient temperature. The warmer it is, the faster they burn fat. If the monarchs lose too much fat, they cannot make it back to the United States in the spring. On the other hand, temperatures below freezing can kill them. The old-growth fir forests of the Transvolcanic Belt provide just the right climate for the monarchs. Temperatures inside the stands are cold enough to keep the butterflies in a semistupor for most of the winter (thereby minimizing the rate at which they burn fat) but rarely fall low enough to kill them. The monarchs show a particular fondness for forests on southwest-facing slopes, probably because those slopes tend to be the warmest.

There is a risk to being too lethargic, however. First, wind or rain can knock monarchs to the ground. Unless the butterflies are able to fly back up into the trees, they will perish. Second, the monarchs must drink water from time to time to avoid dehydration; this requires them to fly to a nearby stream or pond. Fortunately, temperatures within the fir forests are high enough on some winter days to permit the butterflies to fly around, giving the insects an opportunity to reposition themselves (if they have been blown off their perches) and find water.[13]

With the discovery of the Mexican wintering grounds, some of the other mysteries associated with the monarch's annual trek fell into place. Butterfly collectors, for example, had long noticed that the first monarchs to appear in the southeastern United States in spring were worn and tattered individuals. A few weeks later, these individuals were replaced by fresher, brighter individuals. Farther north and later in the spring, the collectors were finding a mix of very worn butterflies, slightly worn butterflies, and freshly emerged ones, too. With the discovery of the wintering grounds, all this began to make sense. The monarchs, it seems, engage in a remarkable, intergenerational migration.[14]

As spring arrives in the Transvolcanic Belt, the monarchs become increasingly active. By mid-February, a few of the butterflies have begun mating; by the beginning of March, the butterflies are engaged in a veritable mating frenzy as they begin their northward migration. An army of pregnant females (along with males) fly northward, reaching the southeastern United States in March. They lay their eggs and continue heading north. Old and weak, most do not get very far. The eggs they lay, however, develop into caterpillars, the caterpillars form pupae, and out of the pupae emerge the next generation of monarchs. These individuals continue the journey by migrating north and laying their eggs on milkweed plants along the way. From this first-born generation arises a second generation of monarchs, and they, too, continue the northward trek into New England and southern Canada, laying eggs en route. It is the progeny of these individuals—the great-

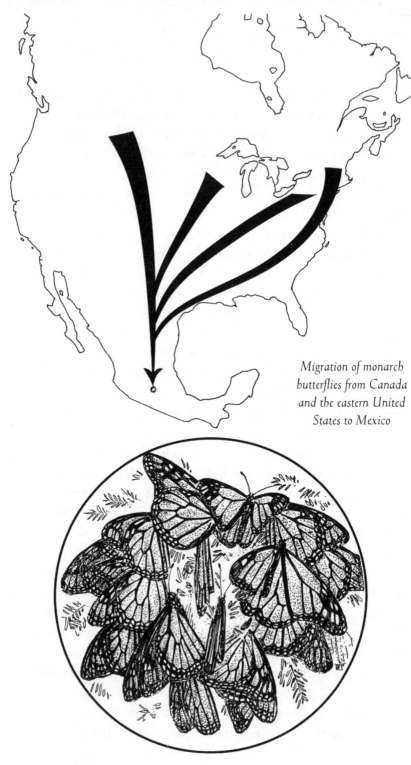

Migration of monarch
butterflies from Canada
and the eastern United
States to Mexico

and great-great grandchildren of the individuals that left Mexico—
that now undertake the journey back to Mexico in the fall. Most of
the members of this last generation of "Yankee" and Canadian butterflies are in a state of reproductive dormancy; they will not breed
until after they have spent the winter in Mexico. Thus, the tattered monarchs that first appear in the southeastern United States
in the spring are the returnees from Mexico. The mixture of very
worn, somewhat worn, and fresh individuals that appear later in
the spring farther north consist of the diminishing pool of sur-
vivors from Mexico and a large number of their children.

The northward advance of the monarchs in the spring appears
to be timed to the emergence of milkweeds, the only plants their
caterpillars will eat. Every spring and summer, billions of milkweed
plants sprout up in the eastern United States and Canada. It's a
bonanza of food for monarch caterpillars, with one catch: winter.
Monarchs cannot survive very cold weather for extended periods of
time, not as eggs, caterpillars, pupae, or adults. That makes all
Canada and most of the United States uninhabitable for them for
much of the year. If the butterflies want to take advantage of the
profusion of milkweeds during the warmer months, they must find
a way to exit during the colder months. Hence, every fall, as temperatures begin to drop, the adult insects begin their remarkable
journey to the mountains of Michoacán. Thus, the monarch migration can be seen as yet another example of opportunism, analogous
to bird migration. Like birds, they head north in the spring to take
advantage of an ephemeral food resource, and they leave in the fall.

How exactly the monarchs find their way to one small region of
Mexico continues to intrigue scientists. There are, in fact, two mysteries to be solved: What navigational tricks do the monarchs
employ to reach their winter quarters? And what environmental
cues or genetic program allows them to produce multiple generations of northward-bound butterflies before producing a generation
that reverses course and heads to Mexico?

On the matter of navigational tricks, there is now strong evidence that monarchs use the sun as a compass (as do we, once we learn the sun rises in the east, is directly overhead around noon, and sets in the west).[15] A sun compass depends upon a circadian clock—the animal must be able to account for the movement of the sun over the course of a day. To demonstrate that monarchs indeed have such a clock, researchers brought migrating monarchs into the laboratory and "clock-shifted" them by exposing them to altered light cycles. These butterflies were then placed in a butterfly flight simulator and reexposed to local daylight. The butterflies, suffering from the insect equivalent of jet lag, proceeded to misread the sun's apparent position and to fly in the wrong direction, demonstrating that they do indeed possess an internal sun compass.

Because monarchs appear capable of migrating on cloudy days (when a sun compass would not work) and because they have magnetite particles in their thorax, some biologists have theorized that the butterflies are also capable of navigating via the earth's magnetic fields. Recent experiments, however, have not supported this idea. Finally, migrating monarchs—like dragonflies and birds—concentrate along coastlines and mountain ranges, which could mean they are using major landscape features for orientation.

How the butterflies manage to wait several generations before initiating a southbound migration remains a mystery. Lincoln Brower, the world's authority on the monarch butterfly, has hypothesized that they migrate in an innately specified direction that changes over the course of the year.[16] Specifically, the monarchs depart from Mexico in a northerly direction. An internal clock causes them to shift their route clockwise at a rate of one degree per day. Such a mechanism would, after six months, result in a generation of monarchs that heads south to Mexico. Six months later, those same butterflies would be ready to head north, thereby initiating the cycle all over again. Brower's hypothesis is appealing in its simplicity, but it has not been verified experimentally.[17]

Almost as soon as the monarchs' winter home was discovered, scientists began to express concern for the butterflies' survival. Their reason was simple: the butterflies congregate in a tiny region of Mexico—an area of roughly seventy-five by thirty-five miles—where deforestation is rampant.[18] In 1986, in response to international pressure, the Mexican government banned logging around the key wintering areas and created a forty-five-thousand-acre reserve. But it did not compensate the landowners who lived in the region, leading to simmering resentment over the loss of control over the land and its natural resources. With little effort to enforce the decree, logging continued to occur in the nominally protected sites.[19]

Severe winter storms in 1991–92 and 1995–96 resulted in the deaths of millions of monarchs. Officials in the Mexican government argued the losses were the result of abnormal weather conditions rather than human interference, an argument Lincoln Brower found unpersuasive. He believed logging was the primary culprit. "The mortality rate went up because the forests were thinner. The forest is like a blanket for the monarchs. Now there are holes in the blanket, so they freeze to death."[20] (In fact, the forest is both a blanket and an umbrella. In addition to providing protection from the cold, the fir trees shelter the butterflies from rain. Wet butterflies are much more vulnerable to cold weather than dry ones[21]). Fortunately, the monarch population as a whole demonstrated a remarkable capacity to rebound, and numbers returned to normal within a year or two of the winter catastrophes.[22]

As word of the monarch's winter home spread, growing numbers of tourists from Mexico and around the world have flocked to Angangueo and nearby communities to experience the phenomenon. Butterfly tourism quickly became an important part of the regional economy, with upwards of 150,000 to 250,000 people now visiting the most popular site, El Rosario, each year.[23] But the revenues from tourism have not benefited everyone, least of all many of the farmers and loggers who earn a living from cutting down the trees. They have not stopped logging.

In 2000, the World Wildlife Fund released a study showing that logging had reduced the average size of the protected patches of forest from 5,000 acres to 500 acres in less than thirty years. The authors estimated that the remaining forests would be gone within fifty years if the rate of deforestation were not reduced.[24] That year, in response to mounting criticism, the Mexican government expanded the size of the reserve to more than 132,000 acres. It also agreed to provide compensation to local residents for the lost timber revenues stemming from the expansion of the preserve.

Yet the deforestation continues, usually in the summertime after the tourists have left. It is driven by a growing demand for wood for use in factories and by construction companies, coupled with the poverty afflicting many people in the countryside surrounding the preserves and the deep resentment felt by many over restrictions on the use of their land. More ominously, much of the illegal logging now appears to be the work of heavily armed, well-organized crime syndicates, whose members have taken to ambushing police officers and terrorizing villagers who try to stop them. In an effort to defend the forests, the president of Mexico deployed the army to the region in May 2004. "We are at war," declared Gabriel Mendoza Jiménez, deputy secretary of public security for the state of Michoacán. "This is not only a problem of cops and robbers. This is a fight for civil order over impunity."[25]

The situation is dire but not hopeless. The Mexican government appears to be getting serious about protecting the butterflies. Sawmills have been closed or fined, and loggers have been arrested. More important, perhaps, some of the local residents have begun to speak out in defense of the monarchs, fearing the loss of the butterflies and the loss of the lucrative tourism they attract. In one community, grandmothers and children have set up plastic tents beneath the stately trees in an effort to deter the loggers. Meanwhile, bouts of bad weather continue to kill lots of monarchs, as happened again during the winters of 2001–2002 and 2003–2004. Each time, of course, we hope the butterflies will rebound like they have on other occasions. But with each fir tree that falls to the chain saw, the odds of a recovery are reduced.

U.S. citizens lamenting the plight of monarchs in Mexico would do well to remember that many of the sites along the California coast that are so important to the welfare of western monarchs are themselves far from secure.[26] The majority of these sites consist of stands of Monterey pines and eucalyptus trees on private property. Some are threatened by real-estate development. In other cases, storms and natural mortality have taken a toll on the older trees; unless these sites are reforested, they will degrade to the point of being useless to the butterflies. A few communities have enacted zoning rules to protect key sites; others have passed expensive bond initiatives to purchase critical areas. But in California as in Mexico, constant vigilance will be required to protect the monarch migration.

Moreover, changes in agricultural practices in the United States and Canada pose a threat of a different kind to monarchs.[27] Prior to the arrival of European settlers, the monarch's principal breeding grounds were probably in the eastern Great Plains, where over twenty species of milkweeds provided abundant food for the caterpillars. Elsewhere in the eastern United States and Canada, smaller numbers of monarchs occurred wherever fires, windstorms, or other disturbances created openings in the forest that were colonized by milkweeds. With the spread of modern agriculture, most of the prairies and much of the eastern forest were converted to corn, soybeans, and other crops.

Fortunately for the monarchs, one species of milkweed, *Asclepias syriaca*, proved adept at surviving along the edges of farm fields or even among the row crops. To the farmers, it is an annoying weed; to the monarchs, it is often the only plant in the area suitable for rearing caterpillars. With the creation of genetically modified, herbicide-resistant crops, however, farmers are now able to spray their fields with herbicides more frequently in order to control weeds. Fields treated in this way are virtually devoid of milkweeds. Thus, the millions of acres of herbicide-resistant crops that are planted each year represent a vast, ongoing loss of breeding and stopover habitat for monarchs and other insects.

WHILE THE POTENTIAL DESTRUCTION of the monarch migration worries people around the world, the obliteration of an even greater insect migration in the United States about a century ago generated no alarm whatsoever. In fact, it was cause for celebration. The difference may have something to do with the creature in question, for instead of it being an exquisite orange and black butterfly, it was a locust, and its once-vast swarms inflicted immense damage to farms across the Great Plains and Northern Rockies. The rapid disappearance of the Rocky Mountain locust, *Melanoplus spretus*, remains one of the great mysteries of American wildlife, for it involves the apparent extinction of an animal that once amounted to a Biblical plague.[28]

To be sure, Americans have an impressive record of decimating once-numerous species, the passenger pigeon and bison being prime examples. But to eradicate an insect whose swarms once covered thousands of square miles, a creature that seemed capable of obliterating the crops of entire counties, takes extinction to a whole new level. We will probably never really know what happened to the Rocky Mountain locust—the historical record is too incomplete to provide a definitive answer—but scientists have uncovered some promising leads in recent years.

The name *locust* is given to certain species of grasshoppers that periodically form vast, migratory swarms. Locusts may live for many generations as nonaggregating, nonmigratory insects; this is referred to as their solitary phase. After successive years of good rainfall and moderate weather, their population in a given area will build. When the good times come to a halt (as they inevitably do), locusts pack into the remaining patches of green vegetation. Under these crowded conditions, chemical, visual, and tactile stimuli initi- ate a remarkable transformation.[29] The insects bump up against one another, thereby stimulating special hairs on their hind legs; they smell one another; they see one another. The nymphs then begin to aggregate, moving together as a broad front or "hopper band" across the landscape. As they mature, they grow wings that are slightly different in shape from those of their parents and grand- parents; they may change color; and finally, they come together as a swarm, flying off en masse in search of greener pastures. The locusts have entered their migratory, gregarious phase. It is akin to a Dr. Jekyll/Mr. Hyde transformation that plays out over the course of several insect generations.

The swarming locusts will eventually settle in an area, gorge on the vegetation, mature, and lay eggs in the soil. When all those eggs hatch, the density of nymphs is likely to be very high, prompting them to be migratory as well. Over the course of a few seasons, however, a combination of natural mortality and variation in weather and rainfall whittles down the locust population, until the density falls below a threshold value and the insects no longer follow the Mr. Hyde developmental pathway. Instead, they become nonaggregating and less mobile. Successive generations will remain in the stationary phase until a sequence of favorable years once again produces the high densities of youngsters that trigger the formation of swarms.

Other species of grasshoppers in North America are capable of achieving high densities on occasion, but none comes close to the Rocky Mountain locust. For settlers living in the Great Plains during the nineteenth century, locusts were an unpredictable menace. Years might pass with no sign of them, and then one hot summer day a cloud would appear in the sky, a strange, dark cloud unaccompanied by thunder or lightning. Within minutes, the fields would be covered in flying locusts—millions, even billions, of ravenous insects gnawing on any plant life they could find. The swarms might linger in a given area for a few days or weeks and then, just as quickly as they came, they would disappear, leaving behind devastated farms and frightened settlers.

The largest swarm on record swept through eastern Nebraska in June of 1875. Albert Child, a local physician with an intense interest in meteorology, took it upon himself to measure the swarm. By telegraphing people to the east and west of him, he determined that it consisted of a continuous front of locusts, over 110 miles long, moving northward. The wind was blowing at roughly 10 miles per hour, and the locusts were moving faster than that; he calculated their speed at 15 miles per hour. Multiplying that rate of travel by the number of days it took the swarm to move through southern Nebraska, he estimated a locust swarm roughly 1,800 miles in length and 110 miles in width, or roughly 198,000 square miles. Such a swarm, as University of Wyoming entomologist Jeffrey Lockwood has noted, "would encompass the combined areas of Connecticut, Delaware, Maine, Maryland, Massachusetts, New Hampshire, New Jersey, New York, Pennsylvania, Rhode Island, and Vermont."[30]

The size of this swarm is so much greater than any other locust swarm ever recorded (the runner-up being a 100-square-mile swarm in Kenya in 1954) that one is forced to conclude that Child's calculations were off the mark. But accounts from other settlers leave little doubt that whatever the size of the swarms, they were more than enough to constitute a nightmare for the men and women struggling to make ends meet in the harsh environment of the Great Plains in the late 1800s. A combination of a locust invasion followed by a harsh winter was enough to drive some communities to the brink of starvation. A report on the Rocky Mountain locust prepared by the U.S. Department of the Interior summarized the situation succinctly: "[It has] brought ruin and destitution to thousands of our Western farmers, and it constitutes to-day the greatest obstacle to the settlement of [the] country between Mississippi and the Rocky Mountains."[31]

The farmers first tried chasing the locusts away from their fields, which proved as futile as fighting the dawn. They then realized that the insects were not only eating their crops but also were laying eggs in the soil. If the farmers could target the eggs or the hatchlings, they might be able to short-circuit the invasion. This hope, in turn, led to the invention of strange devices like the Peteler Locust-Crushing Machine and the King Suction Machine, which, via rotating drums, rollers, wheels, and fans, attempted to crush, churn, or suck the hatchlings into oblivion. Still other machines aimed to exterminate the pests by burning them out of the soil. None was successful, except perhaps in lining the pockets of a few inventors and manufacturers. A somewhat more effective strategy was to flood the fields with water, which did in fact destroy the eggs. But water was (and still is) a scarce commodity in much of the Great Plains, making this an impractical option for most farmers.

Overwhelmed, the farmers turned to their county, state, and federal leaders for help, but in an era when relief was not seen as a responsibility of government, their pleas generated more debate and confusion than actual assistance. Eventually, the army was authorized to distribute clothing and food to some of the hardest-hit areas, and some of the states agreed to provide limited cash payments and loans to destitute farmers. But neither the farmers nor their elected leaders knew how to handle the periodic disasters caused by this one tenacious insect species.

And then the locusts disappeared. In the late 1880s and early 1890s, a few scattered swarms appeared in Minnesota, Nebraska, and Iowa, but none thereafter. Because the swarms had always been unpredictable, neither the farmers nor the agricultural experts made much of the locusts' disappearance at first. But as the years went by without a recurrence of the swarms, it became apparent that something had happened to the Rocky Mountain locust. On July 19, 1902, amateur naturalist Norman Criddle collected two Rocky Mountain locusts on his family's estate in Manitoba, the last ones ever seen alive.

Over the years, a number of hypotheses have been offered to explain the disappearance of the Rocky Mountain locust. None is particularly satisfying, and some are downright silly. Take alfalfa, for example. Studies have shown that young locusts fare poorly on a diet of alfalfa; something about the plant prevents them from developing normally. The acreage of alfalfa in the Great Plains grew considerably during the period when the locusts vanished—could this be the cause of their decline? Unlikely. Alfalfa production was simply not widespread enough in the Great Plains and Rocky Mountains to drive the locusts to extinction. Moreover, since alfalfa production requires an immense amount of water, the crop is grown primarily in irrigated valleys; the upland areas therefore would have remained suitable for locusts. At best, the growth of the alfalfa industry might have cut into the locust population, but it could not have driven the species to extinction.

Attention next turned to bison. The locust plagues of the late 1800s corresponded to the demise of the bison herds (see chapter 4). With tens of millions of hungry bison out of the picture, could the locusts have found themselves in an environment with abnormally large amounts of grass, enabling their populations to explode? In other words, were the great locust swarms a result of the extirpation of the bison? This hypothesis flounders for the simple reason that locusts were swarming long before the bison were extirpated. Large aggregations of locusts were reported in the 1830s, for example, at a time when bison were still reasonably numerous (albeit declining). A revisionist version of the bison hypothesis has also been proposed: the demise of the locusts was actually due to the demise of the bison. Under this theory, the

short grass and disturbed soil that resulted from bison grazing created precisely the right habitat for the locusts. Once the bison were
extirpated, the locusts were doomed. This, too, seems far-fetched.
What little we know about the Rocky Mountain locust (based on
observations by entomologists in the late nineteenth century) suggests it was not so picky in terms of its choice of habitat. Moreover,
as settlers drove the bison out, they brought millions of livestock
in. Wouldn't cattle have created suitable habitat for the locusts?
Cattle are not ecologically equivalent to bison—they are less
mobile than bison, and they tend to congregate near water—but
they ought to be similar enough to create the patches of short grass
and disturbed soil the locusts allegedly required.

With neither alfalfa nor bison providing an especially compelling answer, entomologists next turned to the weather. There
had been some harsh blizzards in the northern Rockies and Great
Plains in the 1880s, and perhaps they destroyed the locusts. This,
too, seems far-fetched. Rocky Mountain locusts have been around
for thousands of years; they surely have survived hundreds, even
thousands, of blizzards as fierce as anything they faced in the
1880s. There is no evidence of any major, widespread change in the
climate during the late nineteenth century that could account for
their disappearance.

And there the matter stood, a bizarre natural-history murder
mystery, until Jeffrey Lockwood began investigating the case in
the 1980s and 1990s. Following up on old reports of receding glaciers leaving behind piles of putrefying, dead locusts, Lockwood
bet that errant swarms of Rocky Mountain locusts had from time
to time landed on glaciers and become entombed in the ice. They
might still be there today, frozen in time. He and his students
explored glaciers in the northern Rockies in search of the insects
and finally hit pay dirt in Wyoming's Knife Point Glacier, finding
vast numbers of frozen locusts. Careful examination of the best-
preserved specimens confirmed they were Rocky Mountain

locusts. Lockwood and colleagues also were able to show that some of the locusts trapped in the ice were hundreds of years old, additional proof that the locust swarms of the 1860s and 1870s were not freakish events but rather normal manifestations of the insect's ecology.

More significant, perhaps, than Lockwood's fieldwork is his analysis of the historical literature on the Rocky Mountain locust, for it enabled him to come up with the best explanation to date for the species' demise. The key was to link the biology of locusts with the history of human settlement in the United States.

Different species of locusts occur in Europe, Africa, Asia, Australia, North America, Central America, and South America. Most of them have a core range or recession area, where they occur all the time, typically in the low-density solitary phase. As the population grows and the insects form swarms, they disperse across a much larger area. Thus, the recession area is the source of the swarms, the base of operations. In the case of the Old World desert locust, the recession area encompasses portions of roughly twenty-five countries in northern Africa, the Middle East, and Asia.[32] It is a huge area, so large, in fact, that the insects defy easy control. Individual countries and international organizations like the Food and Agriculture Organization (FAO) have banded together to track, monitor, and attack locust swarms. Akin to a military campaign, scouts are sent out to find the swarms. Based on reconnaissance, planes and vehicles are then dispatched to attack the locusts, usually by spraying the swarms with organophosphate pesticides.[33] A half century of coordinated efforts at locust control has yielded some notable successes—invasions have been dampened and crops have been spared—although there are skeptics who wonder whether the benefits truly justify the costs. But no one, not even the optimists, predicts the imminent extinction of the desert locust. Complete eradication is not even a goal of the campaign.

In contrast, the recession area (called the permanent zone by nineteenth-century entomologists) of the Rocky Mountain locust may have consisted of just a few valleys in the drainages of the upper Missouri and Yellowstone rivers, places more lush than the surrounding arid lands and mountains yet far enough away from the rivers to avoid frequent inundation. These were precisely the sites white settlers sought out for farming as they headed into Montana and Wyoming. A concerted effort by the federal government to lure people to the West by offering them free land (embodied in the famous Homestead Act of 1862), coupled with the discovery of gold, triggered a mass migration of people into the command center of the Rocky Mountain locust. The settlers broke the sod and planted crops, including alfalfa, whose toxic properties for young locusts have already been noted. Settlers also irrigated their fields, which proved lethal to the locust eggs. The introduction of livestock posed an additional problem for the locusts. As cattle overgrazed the vegetation along the banks of streams and rivers, the banks themselves collapsed, causing the river channels to widen, inundating the eggs of more locusts. The result of all these changes was a swift and thorough assault on the stronghold of the Rocky Mountain locust, according to Lockwood.

Could it have been enough to drive the species to extinction? Perhaps. As the red knot and passenger pigeon remind us, a species that aggregates in a small number of places is uniquely vulnerable to extirpation at the hands of humans. And even if the settlement of the Northern Rockies was insufficient to drive the Rocky Mountain locust to extinction, it may have been sufficient to prevent the insects from ever again reaching a high enough density to make that remarkable physiological transformation into the migratory, gregarious phase. My own suspicion, based on little more than faith in the resilience of insects, is that somewhere in the Northern Rockies, in one or two valleys where the native vegetation has not been completely obliterated, Rocky Mountain locusts hang on. In these remnant pockets of grassland, they spend their days munching on plants, mating, laying eggs, and dying. The amount of suitable habitat left is too little to enable the locusts to reach high enough numbers to trigger the metamorphosis that once caused them to amass and take wing, to become the ravenous, living thunderheads that swept across the plains. Farmers, no doubt, must feel a keen sense of relief that such a phenomenon no longer threatens their livelihoods. But what a sight it must have been.

On Land~

CHAPTER 3

In Search of Greener Pastures

"WHERE THE HELL are THEY?" Here we were in the middle of the Serengeti Plain, grasslands stretching to the horizon, and all around us . . . nothing. Just two weeks ago, according to our guide, Amos Urio, this very place had been awash in wildebeest and zebras, hundreds of thousands of them, but a lack of rains had prompted them to move on in search of greener pastures. Now the only creatures in sight were a couple of Thomson's gazelles, nibbling the brown grass and nervously twitching their tails. I repeated my question. "Where are the herds?" Amos, more bemused than perplexed by the situation, replied, "They're gone." Having come all the way from the United States to Tanzania expressly to see the

Serengeti's wildebeest migration, often called the greatest wildlife spectacle on earth, I was hardly in the mood to accept the Zen-like simplicity of his answer. "How can a million wildebeest just disappear?" I asked incredulously. "They've got to be *somewhere*."

One million wildebeest is a lot of animals. But the Serengeti is a big place, and I had just learned rule number 1 about East Africa's legendary mammal migration: it is tied to the rains. The great herds of wildebeest and zebras that trek between southern Kenya and northern Tanzania do so in response to the seasonal rains. Rule number 2 is that it can be surprisingly difficult to find a million missing wildebeest. We spent the rest of the afternoon and all the next day wandering the dusty roads around the southern end of Serengeti National Park, spotting the occasional herd of gazelles and even a few wildebeest, but nothing approaching the numbers I had come to expect from the scholarly literature and decades of *National Geographic* articles.

Late in the afternoon of the second day, however, Amos spoke with some truck drivers who were passing through the park and had seen sizable herds of wildebeest to the north of us. Early the next morning, Mac Hunter and Aram Calhoun (both ecology professors at the University of Maine) and I piled into Amos's jeep and took off in search of them. We headed east to Naabi Hill Gate, one of the park's main entry points, then north on the main road almost to the very core of the park. Turning onto a dirt road, we bounced across several more miles of open country. Gradually, we began to spot wildebeest—a small herd here, a little group there— and then, quite suddenly, we were surrounded by them. From a bluff overlooking a water hole, we watched in amazement as a broad front of wildebeest, stretching as far south as the eye could see, headed our way. They fanned out into dozens of columns, single- and double-file processions of big, slaty animals marching across the dry, brown savanna, reminding me of lines of giant ants. Their pace was steady but unrushed. The bulls gave a low bray, the cows uttered a nasal "noo," and the calves bleated. Repeated over and over by thousands of animals, their calls merged into a continuous background hum that radiated across the savanna.

As groups of wildebeest drew closer to the water hole, they quickened their pace until, with a few hundred yards to go, they broke into a gallop—bulls, cows, and calves charging across the plains, kicking up a immense cloud of dust in a frenzied rush to get to the water. Once there, they jostled and pushed one another for a position along the water's edge, eventually settling down on their haunches in order to drink. A few became stuck in the thick, black mud and flailed mightily before liberating themselves.

A water hole can be a lifesaver to a wildebeest trekking across an arid grassland. But it can also be a death trap. Lions, leopards, and other predators know that water holes concentrate game, and the dense vegetation along the edges provides numerous hiding places. Perhaps for this reason, the wildebeest around the water hole seemed more agitated than relieved. Time and again, for no apparent reason, they lurched back from the water's edge and galloped away, only to regroup and timidly return. I could never determine whether it was a sound or a smell or the glimpse of a shadow that set them off. On several occasions, these minor routs turned into full-fledged stampedes, with hundreds of wildebeest suddenly racing across the grassland. At such times, calves become separated from their mothers. We watched as one such "orphan" chased after its herd, its spindly legs unable to keep up with the galloping adults. After a few hundred yards of futile pursuit, it stopped, turned around, and headed off in the opposite direction, bleating loudly. Lost, confused, and frightened, the calf raced back and forth between random females, all of whom treated it with utter indifference. The four of us kept hoping its bona fide mother would return to reclaim her offspring, but as the minutes passed, this happy outcome seemed less and less likely.

Biologists that we are, Mac, Aram, and I felt obligated to put a number to the herds, to estimate even crudely how many animals we had seen. We tried various approaches, such as counting the animals passing by a fixed point to get a rate of passage and then multiplying that rate by time, or estimating the density of animals in our field of view and multiplying by the area we were surveying, but we felt confident about none of them. Those who do this sort of thing for a living use airplanes to count the herds; stuck on the ground, we could only guess. By the end of the day, we agreed that we had seen at least one hundred thousand wildebeest, perhaps significantly more, and yet this total, representing one of the most extraordinary days any of us had ever experienced in the field, constituted less than 10 percent of the total number of wildebeest that would eventually head north.

Africa, in fact, is full of mammal migrations. Throughout the continent, there are populations of wildebeest, antelopes, elephants, zebras, and other species that regularly migrate between wet- and dry-season ranges, or between high and low elevations, or from one range to the next. Protecting these herds in the face of an expanding human population with its growing demands for food, forage, and other natural resources is one of the most difficult conservation challenges facing Africans. As we shall see, of the three greatest mammal migrations that once occurred on the continent—wildebeest in the Serengeti, springbok in South Africa, and white-eared kob in the Sudan—only the Serengeti herds live in a reasonably well-protected area, and even their future is by no means secure.

Wildebeest are by no means the only migratory mammals in the Serengeti, although with roughly 1.3 million individuals on the move, they are by far the most numerous. Also migrating are nearly 200,000 plains zebras and over 350,000 Thomson's gazelles. Together they constitute the greatest migration of large mammals left on earth.

Although the three species vary somewhat in the timing and

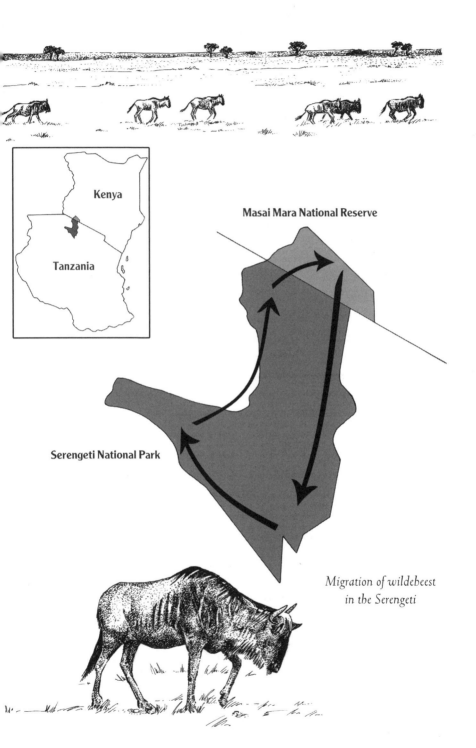

Kenya

Tanzania

Masai Mara National Reserve

Serengeti National Park

Migration of wildebeest in the Serengeti

paths of their migrations, they follow a similar pattern determined by annual rainfall. In the Serengeti, there are typically two periods of rainfall per year: the so-called long rains, which occur in March, April, and May, and the more ephemeral short rains, which occur in November and December. The southeastern part in Tanzania receives significantly less rainfall than the northwestern part near the Kenya-Tanzania border. Moreover, a hard soil pan just below the surface in the southeastern part limits the ability of plants there to put down deep roots. Although the northwestern sector is dom-inated by acacia woodlands, the southeastern part thus features vast plains nearly devoid of trees.

From roughly December to June, the migrants use the Serengeti Plain toward the southern end of Serengeti National Park, Tanzania. Here they graze in large herds, and the females give birth to their calves. In late May and June, the wildebeest move northwest to an area of the park called the western corridor. Then, as the dry season progresses, they head farther north, usually crossing over the border into Kenya's Masai Mara National Reserve. They may spend several months in this woodier environ-ment, grazing amid the thorny acacias. Come November or December, they will reverse course and head back to the plains, thereby completing a round-trip journey of nearly five hundred miles. The zebras tend to move directly north from the plains as they head toward the Masai Mara. The little Thomson's gazelles—or tommys, as they are affectionately known—prefer not to stray as far from the plains as the other two; instead they migrate from the plains to adjacent savannas.

Although the pattern of the migration and its connection to rainfall have been well known since at least the 1970s, the reasons for it remain somewhat mysterious.[1] That the herds would leave the Serengeti Plain in the dry season makes perfect sense. The grass there stops growing soon after the rains cease, quickly turn-ing the plains into a dry stubble that cannot support a million or more big grazers. The wildebeest and zebras then head northwest, to the region of higher rainfall where there is still plenty of grass amid the acacia woodlands. The puzzle is why they ever leave these woodlands, especially after the rains have started and the amount of green forage is increasing. Why not simply stay there all year?

Ecologists have proposed a variety of explanations to explain why the animals trek to the plains each year: they leave the woodlands to escape the hoards of tsetse flies (a perfectly reasonable hypothesis to anyone who has had the misfortune of encountering a swarm of the little bloodsuckers); to escape predators lurking in the ranker vegetation of the woodlands, especially when it is time to give birth; to avoid the soggy ground of the woodlands because wildebeest, at least, can develop foot diseases under such conditions; or to obtain more nutritious forage on the plains.

Of these hypotheses, the one with the strongest evidence in its favor relates to the quality of the forage on the plains. Recent studies have shown that the grass on the Serengeti Plain contains higher concentrations of protein than grass in the northwestern woodlands. This difference could be important to a lactating wildebeest or zebra. Even more striking, however, are the differences in the amounts of calcium and phosphorus available to grazers in the two habitats. Concentrations of calcium in grass samples from the Serengeti Plain are, on average, 40 percent greater than in samples from the northwestern woodlands; phosphorus concentrations in the plains are double the values found in the woodlands. The calcium difference may not be of much significance, as even the reduced levels found in the woodlands should be enough to sustain the animals. But the concentration of phosphorus in the woodland forage appears to be below the level a lactating female wildebeest requires. (Conversely, phosphorus levels on the grasslands are well above that threshold.) In domestic livestock, phosphorus deficiencies have been linked to reduced fertility and milk yield, a loss of appetite, diminished rates of growth, bone and tooth abnormalities, and increased mortality rates; similar problems presumably would affect wild grazers, too.

Thus, it appears as though the Serengeti migration is driven by the need to acquire particular nutrients. Wildebeest, zebras, and gazelles march from the dry-season range to the wet-season range in order to find protein-rich grass with adequate concentrations of phosphorus and (perhaps) other minerals. Their journey, like the annual migration of birds from tropical to temperate latitudes, is an example of ecological opportunism: for much of the year, the Serengeti Plain is hot, dry, and uninviting; but when the long rains arrive, it briefly becomes a lush, nutritious pasture, luring nearly two million wildebeest, zebras, and gazelles. The herds could not survive there year-round; nor, apparently, can they subsist entirely on the less nutritious forage of their dry-season range in the north-western portion of the ecosystem. But by commuting between the two places, an immense population of grazers can sustain itself.

This leaves one small but important bookkeeping detail: in the western and northern parts of the Serengeti ecosystem live resident herds of wildebeest and zebras, consisting of several thousand non-migratory individuals. These animals stay in the savannas and woodlands throughout the year. If the forage in these places lacks sufficient concentrations of essential nutrients to sustain lactating mothers and their calves, then how do these resident herds persist? The answer, according some scientists, may lie in the existence of scattered places across the Serengeti where the soil contains higher-than-usual mineral concentrations. At such sites, the forage is of sufficient nutritional quality to sustain the animals year-round, but the sites themselves are in aggregate too small to sustain the much larger migratory herds.

COMPARED WITH, SAY, a skein of geese winging its way across the autumn sky in a precise V formation, the march of grazers across the Serengeti seems disorganized, unplanned, and at times even chaotic. But, in fact, years of careful observation by Sam McNaughton, Anthony Sinclair, Richard Bell, and other scientists have shown that there is a definite order to the migration of the zebras, wildebeest, and Thomson's gazelles.[2]

This order is dictated in part by the tenfold difference in body size among the migrants, from sixty-pound Thomson's gazelles to six-hundred-pound zebras, and in part by differences in digestive physiology. Not surprisingly, a zebra consumes a lot more food than a gazelle, and unlike the gazelle it can survive on a low-quality diet of grass that is high in fiber and low in proteins and nutrients. The zebra compensates for the poor quality of the food by consuming and processing a lot of it, essentially extracting a small amount of nutrition from a large volume of roughage. The gazelle, on the other hand, consumes much less forage than the zebra, but what it does eat is of higher quality, consisting mostly of tender, young shoots. A wildebeest is almost as big as a zebra, but, unlike a zebra, it is a ruminant. Ruminants have chambered stomachs that enable them to digest their food more thoroughly, thereby extracting a higher proportion of the protein locked within the plant tissue. However, because this slows the rate of digestion, a wildebeest cannot process as much food as a zebra and therefore requires slightly higher-quality forage (but not as high quality as what a gazelle needs).

Scientists have discovered that as the dry season begins, the zebras leave the Serengeti Plain ahead of the wildebeest and Thomson's gazelles. Moving out of the plains, they encounter the tall, rank grass that has grown up in their absence, much of it now dead and brown. Zebras are quite capable of subsisting on this Serengeti "junk food." They consume the coarse top stems and move on. By removing the long, coarse grass, the zebras inadvertently expose the more nutritious leaves and younger stems, which the wildebeest, following in the zebras wake, will eat. Over the course of a few days, the wildebeest reduce the grass to stubble and bare earth, and once they have done so, they depart. The land then undergoes a rather remarkable transformation as green shoots emerge from the soil, their growth prompted by the removal of the rank vegetation and the deposition of immense amounts of wildebeest dung and urine. Within a few weeks, what had looked like a wasteland after the wildebeest's departure is now a verdant lawn, consisting largely of young shoots rich in protein and nutrients. The tommys are drawn to these green areas, which provide them with an abundance of high-quality food.

Ecologists refer to this phenomenon as a grazing succession. Zebras alter the vegetation in such a way as to foster conditions favorable to wildebeest, and wildebeest do so in ways that benefit Thomson's gazelles. However, none of these species is entirely dependent upon the others. Without zebras, there would still be wildebeest in the Serengeti, and without wildebeest, there would still be Thomson's gazelles. But because these animals interact in this way (a result of natural selection), the Serengeti is able to support a greater abundance of grazing mammals than would otherwise be possible.

A question that quickly springs to mind is how the grass is able to withstand the assaults of the grazing mammals (not to mention the countless insects that, in aggregate, probably consume at least as much foliage as the mammals). In fact, the grassland flora is superbly adapted to handle the pressure, even to the point of prospering from it. In many types of grasses, the point of tissue growth, the apical meristem, is located underground. Provided the grazer simply snips off the aboveground growth (as opposed to pulling up the plant by its roots), the plant will send up another shoot. Studies in East Africa have shown that grazing by wildebeest, zebras, and other mammals actually increases the vegetative productivity of the grasslands by removing the senescent growth and causing the plants to send up fresh shoots. Grazers also break up the soil with their hooves and redistribute nutrients via their urine and dung.

Grazing, in short, increases the productivity of the East African grasslands and woodlands. The long coevolution of Africa's grass and grazers has produced a system in which the plants prosper from being eaten and the grazers are able to take advantage of each other's presence. The end result is a wildlife spectacle unlike any other in the world today, unrivaled in terms of the numbers and diversity of large mammals.

The migration of the grazers affects not only their "prey"— grass—but also their predators. In the Masai Mara, for example, spotted hyenas opportunistically switch their diets to wildebeest when the herds move into the area during the dry season. (At other times of the year, hyenas prey on other animals, chiefly gazelles.) The Masai Mara lions appear to adjust their breeding cycle to the migration, and a large proportion of their cubs are born from March through June. By the time the wildebeest herds arrive in July, August, and September, the cubs have become hungry subadults; thus, the pride is able to take advantage of the abundant wildebeest at a time when its need for food is great.

Despite the obvious benefit of having an abundant food source, none of the predators appears to migrate with the herds. This is probably a concession to the special challenge carnivores face in raising their young. Young lions, cheetahs, hyenas, and other carnivores are largely immobile and helpless for the first few months of their lives. Their parents are therefore tied to a particular area or home range until their offspring are old enough to move around. Young wildebeest, zebras, and gazelles, on the other hand, are remarkably mobile within days of being born. Because their young can (usually) keep up with the mothers, the grazers are able to migrate.

Humans, too, have been a part of this picture for hundreds of thousands of years. East Africa is home to the earliest hominid fossils, and it appears likely that our ancestors evolved in a savanna environment not all that different from today's Serengeti. The climate was seasonal, alternating between wet and dry seasons, so presumably the ancestral wildebeest were migratory for the same reasons today's herds are. The arrival of the herds must have marked a glorious time of the year for the early hominids, providing them with an abundance of carcasses to scavenge and wobbly-legged newborn calves to kill. Conversely, the out-migration must have ushered in a period of deprivation and hunger. Yet by becoming bipedal, these hominids could do something no other predators could: they could, quite literally, pick up their offspring and follow the herds. Indeed, some anthropologists speculate that selection for mobility to follow the herds helped drive the evolution of bipedalism in our ancestors (although one can imagine other drivers as well).3 Regardless of its origins, the ability to move around on two legs, thereby freeing up two arms, gave humans a flexibility that no other predators enjoyed.

The wildebeest and zebras were an immensely valuable resource to early humans, but wild animals, however abundant, are less tractable than domestic ones such as cattle, sheep, and goats (none of which occurs naturally in East Africa). Thus, when traders brought cattle, goats, and sheep to East Africa some two thousand years ago, many people switched from being hunter-gatherers to pastoralists.4 They moved their herds across the land in search of forage and water. For the past several centuries, the Maasai have been the dominant pastoralists in East Africa. Traditionally, they have followed a migratory strategy not too different from that of the wildebeest, keeping their cattle and goats in the southern plains during the wet season and then moving them northward during the dry season. The Maasai tend to avoid areas where wildebeest are congregating, partly because the wildebeest herds consume most of the forage (the Koyake Maasai refer to the annual wildebeest migration as their "yearly famine") and partly because the Maasai worry that wildebeest will transmit diseases to their cattle.5 Ironically, however, it was the cattle that ended up transmitting diseases to the wildebeest and other wild animals, leading to major declines in wildlife populations across much of the Serengeti at the beginning of the twentieth century.

The disease in question was rinderpest, a viral disease related to measles. It primarily attacks members of the cattle family (Bovidae), a group that includes wildebeest, antelopes, and African buffalo as well as the domestic cow.[6] Symptoms of rinderpest include fever, ulcers in and around the animal's mouth, nasal discharge, and diarrhea. A high proportion of animals die from the more virulent strains of the disease. Biologists trace rinderpest's arrival in sub-Saharan Africa to the accidental introduction of a few infected cattle into the Horn of Africa in the late nineteenth century. East Africa's wildebeest, buffalo, and antelopes clearly lacked any immunity to rinderpest, and the disease swept across the continent with a fury, decimating wildlife as well as livestock.

A rinderpest vaccine for livestock was developed in the 1950s, and once the cattle were inoculated, pandemics in the Serengeti quickly subsided. Wildlife populations rebounded. Wildebeest, for example, jumped from a mere 220,000 individuals in 1961 to 1.4 million by 1975. (That rinderpest could be suppressed by vaccinating just the livestock suggests that wildlife do not act as a long-term reservoir of the virus, a lucky break for conservationists.) Not surprisingly, this increase led to conflicts between people and wildlife as the Maasai found themselves competing with an expanding wildebeest population for forage.

In recent years, however, growing numbers of Maasai have become sedentary, abandoning their pastoralist traditions in favor of farming. Far from reducing competition between people and wildlife, this transition has resulted in the conversion of thousands of acres of wildlife habitat outside the reserves into cropland. For example, the area under wheat cultivation in a band of ranches to the east of the Masai Mara National Reserve jumped from 12,000 acres in 1975 to over 120,000 acres in 1995.[7]

Two factors in particular appear to be pushing the Maasai to forsake pastoralism in favor of agriculture. First, at least in the short term, farming is more profitable than pastoralism. According to one study, whereas livestock generate $5 to $13 per acre, farming yields anywhere from $50 to $100 per acre.[8] Even tourism, often touted as the economic savior of conservation in East Africa, cannot compete against farming on a per-acre basis; its returns are only marginally higher than those from livestock. (Whether farming in these semiarid lands is sustainable over the long term is debatable, however. Just outside the Serengeti ecosystem, in the Simanjiro plains of northern Tanzania, farmers can expect a good harvest only four years out of every ten. Consequently, they tend to deplete the soil and move on.[9])

Second, the Maasai population has been growing rapidly. One study placed the current growth rate at nearly 4 percent per year, which translates to a doubling of the human population every eighteen years.[10] Because wealth in a pastoralist society is measured in terms of livestock, a growing Maasai population can sustain its standard of living only by increasing its number of cattle, goats, and other domestic animals. This, in turn, requires the pastoralists to find more land for grazing, since each square mile of savanna will support only so many cattle or goats. But with more and more of the savanna being turned into farmland (both by the Maasai and by other tribes), the amount of grazing land is actually decreasing. This leaves the pastoralists with two alternatives: they can overstock the remaining grazing land, which is simply not sustainable over the long term, or they can reduce their herds, resulting in fewer cattle per person and therefore a lower standard of living. With neither alternative offering much of a future, it's easy to see why growing numbers of Maasai are abandoning pastoralism and going into farming and other professions.

To make matters worse, there is something of a positive feedback loop in all this. As more of the land is subdivided and converted to farms, it becomes increasingly difficult for the remaining pastoralists to eke out a living, thereby pushing them to jump aboard the agricultural bandwagon. During the 1970s, for example, the Koyake Maasai and the neighboring Siria Maasai had reciprocal grazing agreements, which allowed each group to use the other's land when necessary. Today, with less grazing land available, such agreements are no longer easy to negotiate because each community needs almost all the land it has just to maintain its own herds.[11] A growing human population outside the game reserves has also resulted in increased amounts of poaching of wildlife. Using snares, pitfall traps, and (to a minor extent) guns, poachers annually harvest tens of thousands of mammals from the Serengeti reserves and surrounding areas, largely for meat.

Thus far, this combined one-two punch of habitat conversion and poaching has not had a major effect on the migratory herds of wildebeest and zebras, but it has been devastating to the resident (nonmigratory) herds. Around the Masai Mara National Reserve at the northern end of the Serengeti ecosystem, for example, resident wildlife populations have declined by over 70 percent since 1980. Resident wildebeest dropped from one hundred thousand individuals to just twenty thousand today, and similar declines have been recorded among nonmigratory zebras, buffalo, giraffes, and gazelles.[12] Opinions differ as to whether habitat conversion or poaching is the bigger culprit, but few conservationists would deny the need to bring both under control if the reserve is to retain its rich array of large mammals.[13]

Why the migratory herds haven't suffered as much as the resident herds is an intriguing question. Part of the answer may be that the boundaries of the two main reserves in the Serengeti—Tanzania's Serengeti National Park and Kenya's Masai Mara National Reserve—were adjusted several times to protect most of the key areas used by the migratory herds. Therefore, the rampant habitat destruction now occurring outside the reserves has had less of an effect on the migrants than on the residents. It's worth noting, however, that the migratory routes of the wildebeest and zebras vary from year to year and even change over time.[14] What appears to be a properly situated reserve system today may prove inadequate ten, twenty, or fifty years from now. And if the areas surrounding the current reserves have been turned into farms and settlements, there may be nothing the migratory wildlife—or its protectors—can do to remedy the situation. For the time being, the greatest mammal migration on earth seems relatively secure, but whether it remains so in the coming decades is anyone's guess.

TO GAIN A SENSE OF what East Africa could look like in another fifty years, one need only travel to South Africa, the wealthiest and most developed sub-Saharan nation. A drive through the South African countryside reminds me of a trip to Iowa: as far as the eye can see, the native grasslands have been converted to agriculture, producing miles upon miles of fenced and manicured farmland, devoid of large mammals other than cattle and sheep.

Several years ago, South African ornithologist Callan Cohen and I were driving past the town of Swellendam, approximately 120 miles east of Cape Town, when we passed a sign for Bontebok National Park. The word *bontebok* rang a bell; I recalled reading somewhere that it referred to one of South Africa's rarest and most beautiful antelopes. "Do they really have bontebok there?" I asked. "Of course," replied Cohen. "Would you like to have a look?" We headed down a gravel road, through gently rolling hills, to the entrance of the park.

In its own quirky way, Bontebok National Park embodies both the triumph and the failure of wildlife conservation in South Africa. By the early 1800s, the once-numerous bontebok had been driven to the brink of extinction. Recognizing that the species was all but finished, a few conscientious landowners in the cape region set aside portions of their land to form a temporary sanctuary for the beleaguered animals. In 1931, the first Bontebok National Park was proclaimed near the town of Bredasdorp; the park subsequently was moved to its current location, which is thought to contain superior habitat for the antelope. It is a very small park—less than fourteen square miles—and it is completely enclosed by a chain-link fence. Inside, the bontebok enjoy a relatively secure, predator-free existence. Park managers periodically cull the herds to keep the number of bontebok close to two hundred.[15]

Shortly after passing through the park gate, we came upon our first group of bontebok, magnificent purplish-brown antelopes with sinuous horns and bold white patches on their faces, buttocks, and bellies. I found myself silently thanking the farmers who had the foresight nearly two centuries ago to preserve the species. Yet at the same time, I found the whole experience of seeing "wild" bontebok strangely unsatisfying. The place seemed more like a zoo than a park and the bontebok more like cattle than wild animals. They face no enemies, their numbers are regulated by vigilant wildlife biologists, and their movement is constrained by fencing. But deep within the bontebok genome, there remains an urge to wander. According to African mammal expert Jonathan Kingdon, bontebok in enclosures still retain "the habit of circulating around their available range in loose herds," a relictual behavior, perhaps, from a time when the cape was a wilder place and herds of bontebok roamed across the fynbos.[16]

The bontebok, in fact, has done rather well compared with some of the other mammals that shared its habitat. The quagga, a stunning zebra with chocolate-colored hindquarters that once occurred around Cape Town and the Karoo, was driven to extinction in the nineteenth century by habitat loss and overhunting. The cape lion, renowned for its luxuriant dark mane, vanished around the same time. Herds of black wildebeest formerly migrated between the Karoo shrublands and the temperate grasslands of the Highveld. They too appeared headed for extinction in the nineteenth century until a few conscientious landowners chose to protect some of the dwindling herds. Like the bontebok, the black wildebeest persists today as a collection of semidomesticated herds in various reserves and ranches.

On that same trip, as we headed north from the cape region into the Karoo, we stopped to admire a small group of antelope grazing in the fields. They were springbok (also known as springbuck), similar in appearance to the Thomson's gazelles of East Africa. As we watched them, Cohen remarked that it was hard to believe that only a century or two earlier hundreds of thousands, if not millions, of springbok regularly migrated through the Karoo. It was my first clue that South Africa once hosted a mammal migration that rivaled what we see today in the Serengeti, although it involved primarily one species.[17]

Many details about the springbok migration will forever be a mystery inasmuch as the phenomenon was snuffed out before scientists had much of a chance to study it. The second- and third-hand accounts of the great migrations of the nineteenth century describe the march of the antelopes in terms usually reserved for locusts and other biblical plagues. An especially vivid description comes from the writer Lawrence Green, who interviewed Gert van der Merwe, a Karoo rancher, about his encounters with springbok in the late nineteenth century.[18]

According to Green, van der Merwe and his family were traveling by oxcart along the hard, dry bed of the Molopo River when their Bushman guide warned them that a springbok migration was fast approaching. Van der Merwe and family drove the oxcart onto higher ground and encircled it with piles of brush, which they set on fire to deter the antelope. Their first clue that the antelope were nearby came when hares, jackals, meerkats, mice, and snakes began streaming past the oxcart as though fleeing from some horrible, unseen danger. Then they heard a faint drumming sound that grew steadily louder.

The first solid groups of buck swept past on both sides of the hill. After that the streams of springbok were continuous, making for the river and the open country beyond. Then the pressure increased, the buck became more crowded. No longer was it possible for them to swerve aside when they reached the fires and the wagon . . . Some crashed into the wagon and were jammed in the wheels, injured and trampled upon. The wagon became the center of a mass of dead and dying buck . . . But the thorn barrier had broken, and the buck were among the cattle. Before long the terrified, bellowing cattle stampeded and vanished into the dust in the direction of the river. Gert had to let them go. There was only death for anyone who ventured after them among the horns and hooves of the bucks.[19]

Within an hour, most of the herd had passed, but stragglers continued to pour by the van der Merwes well into the night. The next morning, Gert "saw that the landscape, which had been covered with trees of fair sizes, green with food for his cattle, were gaunt stumps and bare branches. The buck had brushed off all herbage in their passing, and splintered the young trees so that they would never grow again."[20] Every gully and stream was filled with the corpses of springbok that had hesitated, tripped, or fallen and had been trampled to death by their peers. All around he could see the corpses of other animals—tortoises, hares, and the like—that had met a similar end.

One suspects that either van der Merwe or Green was exercising more than a little poetic license in describing the events of that day, but enough similar accounts exist to confirm that vast numbers of springbok indeed moved through the Karoo. The size of the herds is subject to debate. Some settlers who witnessed the migration swore there were millions of animals; others believed there were "only" hundreds of thousands. Regardless of the precise number, the springbok migration was unquestionably one of the continent's greatest wildlife spectacles. Yet it essentially stopped at the end of the nineteenth century, with the last major movement occurring in 1896. A witness to that last migration estimated that no fewer than a half million springbok were spread out across a plain, "distributed in one unbroken mass over the whole expanse . . . giving quite a whitish tint to the veld, almost as though there had been a very light fall of snow."[21] In subsequent years (including right up to 1990), there would be occasional movements of springbok, but those events were measured in the hundreds or thousands of individuals, rather than in the hundreds of thousands or millions.

By examining historical accounts, we can piece together some details about the springbok migration. First, the frequency of the migrations varied from place to place, from almost annually in the north to once every seven to eight years in the south.[22] Second, the migrations tended to occur during drought years.[23] This finding is especially important because it suggests the springbok were wandering in search of forage. Third, settlers in the nineteenth century asserted that the springbok never returned the same way they came, indicating the migration did not follow a consistent route. The combination of erratic timing, linkage to droughts, and unpredictable routes suggests that the mass movements of springbok did not constitute a classic migration along the lines of the Serengeti wildebeest. Instead, successive years of good rainfall may have caused the springbok population in the Karoo to explode periodically. When the rains failed, as inevitably happens from time to time, the springbok exhausted the available forage and were forced to search for green pastures. As the wandering herds converged on the remaining patches of forage, the aggregations of springbok swelled, ultimately creating the immense herds reported by the early settlers.[24]

If this hypothesis is correct, then from an ecological perspective the great springbok herds were analogous to locust swarms. And like locust swarms, they were deeply hated by farmers. Whenever the springbok herds approached farms and human settlements, the animals were massacred, partly for their flesh, partly for their hides, but mostly because people felt springbok destroyed crops and competed with livestock for forage. "Convoys of wagons, carrying whole families, incepted [sic] the trekbokke [migrating springbok], muzzle-loaders went into action, and one bullet often killed more than one buck."[25] Indeed, an 1886 amendment to South Africa's game laws specifically exempted migrating springbok from protection, largely to mollify ranchers and farmers.[26]

Several explanations, none mutually exclusive, have been put forward to explain the disappearance of the migrations at the start of the twentieth century. Settlers fenced the rangeland, thereby creating barriers to movement. They also introduced large numbers of livestock, which overgrazed the land, thereby reducing its capacity to support wild grazers. Hunting undoubtedly played an important role as well, especially with the advent of breech-loading rifles. Some scientists also point the finger at diseases such as rinderpest, arguing that disease mortality kept springbok populations below the density threshold that precipitated a migration, although a recent review of the historical data found no evidence in support of this particular hypothesis.[27] Instead, it appears as though a combination of fencing, livestock grazing, farming, and shooting simultaneously reduced the numbers of springbok and shrank their world, thereby putting an end to the great migrations.

It's worth emphasizing that the springbok is far from an endangered species in South Africa. It remains a common and widespread animal on farms and ranges, just nowhere near as abundant as it once was. What is gone is not the species but the phenomenon of the species, the spectacle of hundreds of thousands of springbok marching across the Karoo desert, kicking up great clouds of dust, as they wander in search of forage. I have no doubt that few if any of the farmers and ranchers currently living in the Karoo would welcome back the great herds of springbok—the small, manageable herds that live there now are just fine, thank you—but I for one wish I could see what Gert van der Merwe saw. From a safe distance, of course.

SOME TWENTY-FIVE hundred miles to the north, Africa's third great mammal migration seemed destined to go the way of the springbok. The migrant in this case is the white-eared kob, a medium-sized dark antelope handsomely marked with white patches on its ears and around its eyes, muzzle, and throat.[28] During the rainy season, the White Nile overflows its banks, creating vast swamps and seasonal grasslands in an area of southeastern Sudan known as the Sudd. Here the kob gather in July and August to graze and give birth. As the rains fade and the forage withers, the kob begin their trek northward, eventually settling in wetter, greener areas along the Ethiopian border. They will stay in the borderlands until the rains return in April or May, at which point they will once again march to the Sudd, completing a round-trip journey of over four hundred miles.

As with the Serengeti wildebeest, it's easy to understand why kob must leave the Sudd when the rains end. The Sudd lacks a source of permanent water, and kob must drink daily. Moreover, there simply isn't enough forage there in the dry season to sustain the herds. More perplexing is why they bother to leave the swampy areas near the Ethiopian border at the onset of the rainy season, inasmuch as the rains increase the amount of available forage. Perhaps, like the wildebeest, the kob are missing some essential nutrient in their dry-season quarters, forcing them to head south each year.

At least through the early 1980s, a million or more kob made the trek. In 1981, the Sudanese government even created a 5.6-million-acre national park, Boma National Park, to safeguard some of the key habitats of kob and other species. But the long-standing animosity between Sudan's largely Arab population in the north and its non-Arab population in the south erupted (again) into civil war in 1983. Most of the fighting took place in southern Sudan. Over the course of the next two decades, millions of Sudanese were killed, maimed, or displaced by the hostilities. Throughout the region, civil infrastructure, including the national parks, collapsed as rebel and military forces fought for control of the land. There were fears that desperate people, many now armed with weapons, would turn to the wildlife for sustenance. By 2001, reports suggested the white-eared kob population had plunged to 180,000, a decline of more than 80 percent in less than a generation. Then, in 2005, the Sudanese government and the rebels signed a peace agreement, establishing the Government of Southern Sudan. In 2007, scientists were able to return to the region for the first time in a quarter century. They were astounded to find the kob migration largely intact, with 800,000 or more animals on the move. The poachers apparently had been unable to penetrate the remote and swampy Sudd.

Against the immense human tragedy of the Sudanese civil war, the fate of the kob may not seem like a pressing issue. But it's worth remembering that this migratory spectacle is far older than any government, any city, any human institution in that beleaguered region. If the people of Southern Sudan achieve the peace and prosperity they desire, they will surely be grateful they did not sacrifice their country's greatest wildlife phenomenon along the way.

AND WHAT ABOUT THE MULTITUDE of smaller-scale migrations in Africa? Although these movements do not rival what we see in the Serengeti or the Sudan, they are nonetheless crucial to the welfare of the animals that undertake them. For example, south of the Serengeti ecosystem, in Tanzania's Tarangire National Park, many zebras and wildebeest exit the park during the wet season.[29]

As is the case in the Serengeti, their migration is driven by the need to find forage with adequate amounts of phosphorus. Inside the park, phosphorus concentrations are simply too low to support lactating females; outside the park, however, the animals can find areas rich in phosphorus, most notably on the Simanjiro plains to the east. Yet, as the dry season approaches, the large mammals must head back into the park, since it contains the only permanent source of water. Unfortunately, agriculture is claiming more and more of the migratory routes and grazing land outside the park. Of nine key migratory routes identified in Tarangire in the early 1960s, three have been erased by agriculture and settlements and five others are heavily fragmented. In addition, poaching outside the park has taken a heavy toll on some of the animals, most notably the wildebeest. Surveys indicate that Tarangire's wildebeest population has plunged by almost 90 percent over the past decade, largely due to illegal hunting outside the park.[30]

Driving both the poaching and the habitat destruction is that communities outside the park, on whose land the migratory herds wander, receive little economic benefit from the animals. Due to the way Tanzania's wildlife policies are structured, very little of the revenue from tourism in and around Tarangire flows back to the local people. As biologists Charles and Lara Foley, who have studied Tarangire's wildlife for many years, have noted, "Until local communities living adjacent to the Park are able to benefit directly from wildlife-related enterprises—be it from photographic or hunting safaris—the situation of land conversion is unlikely to be reversed."[31]

There are undoubtedly hundreds, if not thousands, of similar small-scale mammal migrations occurring throughout Africa, some well known to scientists and wildlife officials, others known only to local residents, and still others a mystery to all. Many African parks and game reserves were created with incomplete knowledge of these migratory behaviors, resulting in parks that are too small to wholly sustain the wildlife within them. Opportunities to adjust the park boundaries to include migratory pathways and seasonal habitats are disappearing rapidly as a growing (and, one hopes, more prosperous) human population usurps more of the land. Although few of the species are likely to disappear completely, many will decline in numbers or disappear altogether from the smaller, more isolated parks. The end result will be an Africa that looks increasingly like the rest of the world, largely devoid of spectacular mammal migrations, its large mammals restricted to a small number of major parks, the parks themselves encircled by fences, what little is left of the wilderness incarcerated for its own safety.

CHAPTER 4

Where the Buffalo Roamed

By MID-APRIL, winter has released its grip on northeastern Colorado, and the Pawnee National Grassland is no longer a lifeless
stretch of snow, mud, and dead grass. To the contrary, the place is
fresh and green, and the newly arrived birds are engaged in a frenzy
of courtship and homesteading. In this flat, treeless environment,
male birds face a dilemma: how do they make their presence known
to one another and to the females as they go about the business of
setting up territories and attracting mates? For the grassland sparrows—the chestnut-collared longspurs, McCown's longspurs, and
lark buntings—the answer is to take to the air. The males shoot up
into the sky like little champagne corks, singing vigorously as they

climb. At the apex of their flights, they flair their wings and rock back and forth, thereby slowing their descent as they parachute back to earth, singing all the while. Leaning against my car, I look around and see a dozen or more of these little birds popping into the air. This place is magical.

If spring is a time of rebirth for North America's prairies, it is doubly so in this region. The Pawnee National Grassland was stitched together in the wake of the Dust Bowl of the 1930s, when the federal government bought the land from bankrupt farmers. The farmers had intended to break the sod, convert the native grasses to wheat, and then harvest the wheat for a profit—all of which they did, for a while, until drought, overzealous planting, and a stock-market crash turned their dreams into dust. Since that time, wildlife has reclaimed much of the land, as my April visit made clear to me.

Yet despite the abundance of birds, insects, and plants in the springtime prairie, I knew that something was missing. And had I been a better naturalist, I might have noticed the evidence of that loss as I wandered around northeastern Colorado: shallow depressions scattered across the plains. These are thought by some biologists to be old bison wallows, places where itchy bison used to roll around in the dirt, enjoying the bovine equivalent of a good scratch. Even today, well over a century after the last bison awkwardly lay down and kicked up a cloud of dust, the wallows persist, although now overgrown with grass—ghostly remnants of the great herds that once filled the prairies.[1]

Despite all that has been written about the loss of the American bison, it remains a poorly understood event. An ecologist might even say it is a story with an ending but no beginning because we know so much more about the demise of the bison—the slaughter of the herds and the efforts to save the last few individuals—than we do about the species in its heyday. How many bison were there before the slaughter began? Where did they go?

What did they do? Questions like these will always be shrouded in mystery, partly because the destruction of the herds happened so quickly and partly because there were no scientists around to study the animals when they were abundant. The best we can do is to piece together the fragmentary accounts of the early explorers and combine them with contemporary studies of remnant bison herds to build a portrait of the presettlement situation.

This we know: prior to the American Revolution, bison occurred over most of North America, from New York State south to Georgia and Mexico, west to the Rocky Mountains (perhaps even to Nevada), and north to Canada's Northwest Territories and Alaska. Their center of abundance, however, was in the grasslands of the midcontinent. Ecologists traditionally divide this region into three sections, distinguished by the height and species composition of the vegetation. The eastern section, stretching from southern Manitoba to southern Texas and from Kentucky to Nebraska, is the tallgrass prairie, which once covered about 148 million acres. To the west, the tallgrass prairie merges into the mixed prairie, a band of drier, shorter grassland that extends from Alberta through the center of the Dakotas, Nebraska, Kansas, and Oklahoma to central Texas. The mixed prairie, in turn, merges into the even drier, even shorter shortgrass prairie, which extends along a north–south axis from the western Dakotas to northern Texas and westward to the Rockies. Prior to human disruption, the combined mixed-grass and shortgrass prairie covered over 400 million acres.

Bison were most abundant in the mixed and shortgrass prairies. Here they gathered in large herds during the summer, when a combination of warm weather, sunlight, and rain created a productive pasture. During the winter months, when the grass-lands were covered in snow, the bison generally split into smaller groups and sought shelter and forage in wooded draws and riparian forests.[2] The animals were migratory in two contexts. First, they moved from wooded areas onto the grasslands with the advent of warm weather. Second, once the herds formed, they moved across the grasslands over the course of the spring, summer, and fall as they depleted the forage in a given area. How far they roamed during their migrations is, unfortunately, a mystery. Naturalist Ernest Thompson Seton, writing in 1909, concluded that the herds migrated "from 300 to 400 miles in spring, and as far southward again in autumn, but that the regularity of this movement was often much obscured by temporary changes of direction to meet changes of weather, to visit well-known pastures, to seek good crossings of rivers or mountains, or to avoid hostile camps and places of evil memories."[3] He may well have been right, but he provided little hard evidence in support of this statement.

Nor do we know precisely how many bison lived in the central grasslands prior to the arrival of white settlers. Early explorers were awestruck by the numbers they encountered. They describe a sea of bison moving across the plains at certain times of the year, riverboats blocked by immense herds crossing the water, and locomotives derailed by belligerent bulls. By one estimate, a single herd sighted in the vicinity of the Arkansas River may have contained more than four million animals. Fascinated by this question, Seton used a variety of approaches to estimate the original abundance of bison: he divided the area of the continent by the acreage necessary to support a single bison (allowing for different densities of bison in different habitats); he extrapolated the number of bison from the number of horses, cows, and sheep that ranchers were able to keep on the land after the bison had been eradicated; and he used the estimate of four million bison along the Arkansas River as the basis for an overall density estimate for the continent.[4] In the end he concluded that somewhere between 50 and 75 million bison once occupied North America, with the vast majority (up to 70 million) occurring in the central plains.

More recently, some scientists have challenged the accuracy of Seton's estimate, arguing that he grossly overestimated the carrying capacity of the range. Yet even these critics concede that millions, perhaps even tens of millions, of bison once roamed the central plains. Even at those lower numbers, it would have constituted the greatest aggregation of large mammals on earth.

Nor were bison the only large mammals on the prairies. The number of pronghorn probably rivaled the number of bison. Large numbers of elk also inhabited the grasslands. And topping all these species in sheer abundance, including the bison, were the prairie dogs. Four species of prairie dogs occur in North America: black-tailed, white-tailed, Gunnison, and Utah. Of these, the black-tailed prairie dog was by far the most numerous, occupying much of the central plains from Canada to Mexico. At an average density of four to twenty-two prairie dogs per acre, the total blacktail population must have numbered in the hundreds of millions, if not billions.[5]

Although one might not expect bison, pronghorn, and prairie dogs to have much to do with one another, given how different they are, the three species have a surprisingly complex relationship, somewhat analogous to the relationship between zebras, wildebeest, and Thomson's gazelles in the Serengeti.[6] Prairie dogs constantly manipulate the soil and vegetation around their burrows, pushing up mounds of dirt and clipping and eating the grass surrounding their towns. Because of these activities, other types of plants, chiefly forbs and shrubs, are able to gain a foothold around the dog towns, much to the delight of the pronghorn, which prefer forbs to grasses and consequently spend a great deal of time browsing on forbs near the centers of dog towns.[7]

Along the edges of their colonies, prairie dogs clip the tall grass to prevent predators from sneaking up on them undetected. By doing so, they promote the growth of tender, young grass shoots, which in turn attract bison. In one study in South Dakota, biologists found that bison spent approximately 40 percent of their time on prairie dog towns, especially along the edges, even though dog towns made up only 12 percent of the landscape. The presence of the grazing bison probably redounds to the benefit of the prairie dogs as well, if the bison help to keep the grass around the dog towns short (for predator avoidance) and nutritious. In this manner, the grazing mammals create a mosaic of habitats (defined by the height and species composition of the vegetation) based on where they go and what they eat.

Fire, too, contributes to the diversity of habitats within prairies by removing most of the aboveground vegetation and promoting a flush of new growth. Although ecologists have long recognized the critical role of fire in maintaining prairie ecosystems, they are less certain about where, when, and how such fires occurred.[8] Part of the problem is that grasslands, unlike forests, retain no lasting evidence of a fire. In the case of forests, ecologists can use the scars left on tree rings to determine the frequency of wildfires; no such evidence persists in grasslands. Moreover, American Indians set fire to the grasslands for thousands of years prior to the arrival of white settlers, and there is no easy way to separate their fires from those caused by lightning. All that said, ecologists are in general agreement that fires were a frequent event in the central grasslands. Fire return intervals may have been as short as two to five years in the tallgrass prairies and every four to five years in the mixed-grass prairies.

Thus, even though the Serengeti and central grasslands of North America are a world apart and have almost no species in common, they operate under somewhat similar rules: a combination of fire and grazing creates a diversity of habitats within the grasslands. This habitat mosaic, in turn, enables the different species to find the food, shelter, and other elements they need to survive. In both cases, the movements of the large grazers are dictated by the availability of forage, which, in turn, is related to seasonal patterns of precipitation.

There are, however, some important differences between the two systems. The Serengeti represents a grazing succession. Zebras and wildebeest generally move ahead of gazelles; grazing of the taller, ranker vegetation by zebras and wildebeest prompts the growth of the younger, more nutritious forage the gazelles require. In the case of North America, bison, pronghorn, and prairie dogs occur contemporaneously, each species taking advantage of the habitats the others have created or maintained. There is no evidence of a grazing succession in North America's grasslands akin to what is seen in Africa. However, we ought to be somewhat cautious about drawing conclusions about the Great Plains, given how little we know about this region in its undisturbed state. It is certainly possible, for example, that long before settlers decimated the bison, large numbers of pronghorn or elk followed the great bison herds in order to take advantage of the younger, more nutritious grass and forbs that would have sprouted after the bison had consumed the taller, ranker vegetation.9

Whether in Africa or North America, the presence of so many tasty herbivores inevitably attracts predators. In the case of North America's grasslands, the dominant nonhuman predators for the past eight thousand to ten thousand years have been gray wolves and grizzly bears. If today we tend to think of these animals as inhabitants of rugged mountains and remote forests, it is because white settlers quickly exterminated them from all but the most remote areas. But to explorers and settlers at the start of the nineteenth century, grizzlies were a familiar and frightening sight in the foothills and river valleys of the Rocky Mountains, and wolves were often encountered on the open plains.

For Indians living along the fringes of the central plains, bison were an important source of food. However, the evolution of Indian cultures wholly dedicated to hunting bison did not evolve until the eighteenth century. Prior to that time, the migratory behavior of the bison meant the tribes could not depend on them for food throughout the year. The missing ingredient was the horse, which allowed hunters to track the migratory herds. Wild horses disappeared from the New World at the end of the Ice Age and did not reappear until European explorers brought them over in the late fifteenth and early sixteenth centuries. As Spanish colonists occupied the large plateaus of northern Mexico in the late sixteenth century, they began to let their cattle roam the open range. Managing these dispersed herds required lots of horses. By the early seventeenth century, Spanish stockmen had crossed the Rio Grande and were running large herds of cattle and horses in New Mexico, usually with Indian assistance. Eventually, some of these Indians broke away from the Spaniards and rejoined the tribes, bringing with them some horses. In this way, the horse culture spread through the Great Plains, reaching all the tribes between 1650 and 1770.[10]

Equipped with a new mode of transportation, the tribes could now follow the bison herds year-round. As a result, a number of tribes converged on more or less the same lifestyle, adjusting their social organization to match the rhythm of the bison—rejecting permanent settlements in favor of mobility, forming large hunting parties in the summer, when herds of bison were gathered on the shortgrass plains, then breaking into smaller, more dispersed hunting groups during the winter months, when the bison herds tended to fracture into smaller groups and seek shelter in riparian areas and wooded draws.[11] The bison-dependent tribes limited their possessions to what they could carry, and they traded extensively with the sedentary tribes for food items and other goods they could not obtain from hunting. In short, they became as migratory as the bison themselves.

Many tens of thousands of people—primarily Arapahos, Assiniboines, Atsinas, Blackfeet, Cheyennes, Comanches, Crows, Kiowas, and Sioux—adopted this lifestyle, moving deep into the plains in pursuit of bison. Thousands more were opportunistic bison hunters, taking the animals when the herds were in the vicinity of their settlements but not pursuing them year-round.

The question naturally arises, were these tribes hunting bison at a sustainable rate? Without better data on the population sizes of the tribes, the number, age, and sex of the bison they harvested, and the birth and death rates of bison themselves, it is impossible to know. In 1859, one observer estimated that the bison-dependent tribes were taking approximately 450,000 bison per year for their own consumption and for intertribal trade.[12] Depending on how many bison inhabited the plains (a big unknown), such a harvest could have amounted to anywhere from less than 1 percent to approximately 5 percent of the total plains bison population each year. Neither percentage seems particularly high, but without knowing how many bison were perishing due to other causes, such as wolves and blizzards, we cannot place these values in any sort of context.

Whether sustainable or not, the bison-dependent societies were a short-lived phenomenon. With the arrival of white hunters and increased commerce between whites and Indians, bison hunting quickly became bison slaughter. In less than a century, the most abundant large mammal in the world would be brought to the brink of extinction. The destruction of the bison grew out of several factors: a seemingly insatiable desire for their hides; the desire of white settlers to settle the Great Plains; and a belief that if the bison could be eliminated, the Indians could be destroyed or, at the very least, "civilized," which meant putting an end to their nomadic lifestyles and getting them onto reservations.[13]

During the first third of the nineteenth century, the market for bison skins ("robes") grew steadily, probably because other fur bearers, such as beavers, were already suffering from overexploitation. By the 1850s, however, trade in bison robes was already on the downslide, reflecting the overexploitation of the herds. Yet before the robe trade could fade away, the rapid industrial growth in the United States in the 1870s and early 1880s created a new and different market for bison products, this time for their leather, which was used to make belts for industrial machinery.

Factories had been using cowhide for that purpose, but the demand for leather outstripped the domestic supply, forcing manufacturers to import cowhides from Latin America. The continuing demand for leather, coupled with the high cost of importing it, made the bison herds in the West a tempting source. Once a few kinks in the treatment process were worked out, industrialists discovered they could turn bison hides into top-quality industrial leather. Add to this the invention of more accurate and powerful guns plus a growing railroad network across the Great Plains that made shipping easier, and the end of the bison was all but inevitable. As historian Andrew Isenberg has noted, "The hunting of the bison in the 1870s and early 1880s was unquestionably the work of an industrialized society. The western plains became a remote extension of the global industrial economy and an object of its demand for natural resources."[14]

At the same time that bison were being shot to extinction for their hides, they were also coming under fire from ranchers, who believed—justifiably—that the herds would compete with livestock for forage. Still other settlers and government officials felt that eliminating the bison would force the remaining plains Indians to give up their traditional lifestyles and settle on the reservations. Even Columbus Delano, secretary of the interior from 1870 to 1875 and the man nominally charged with safeguarding the nation's wildlife, seemed to welcome the end of the bison, noting that he "would not seriously regret the total disappearance of the buffalo from our western prairies, in its effect on the Indians, regarding it rather as a means of hastening their sense of dependence upon the products of the soil and their own labors."[15]

By the early 1880s, the bison stood on the brink of extinction, reduced to a handful of small herds scattered across the American and Canadian plains. A young Theodore Roosevelt, living on a ranch in North Dakota in the wake of the slaughter, noted, "No sight is more common on the plains than that of a bleached buffalo skull; and their countless numbers attest the abundance of the animal at a time not so very long past . . . A ranchman who at the same time had made a journey of a thousand miles across northern Montana, along the Milk River, told me that, to use his expression, during the whole distance he was never out of sight of a dead buffalo, and never in sight of a live one."[16]

So great had been the carnage that, for a brief period, a thriving industry developed based on gathering bison bones from the grasslands and grinding them up for fertilizer and bone black (a pigment). It took one hundred skeletons to produce a ton of bones, which fetched $4 to $12 in the market. Thousands of tons were shipped to factories annually for a few years. Ironically, many of the people employed to gather the bones were Indians. In essence, they were being paid to harvest the remains of their culture.

If the secretary of the interior seemed unconcerned about the impending extinction of the American bison, others were horrified at the prospect. The close of the nineteenth century was a time of wholesale slaughter not just of bison but of much of American wildlife. From coast to coast, mammals and birds of all shapes and sizes were being killed in unprecedented numbers for personal consumption and commercial sale. Passenger pigeons and Carolina parakeets had all but disappeared; sea otters and northern elephant seals were nearly gone as well; and numerous waterfowl, shorebirds, and wading birds were much reduced in numbers. Add to these developments the loss of the bison, arguably the preeminent symbol of the American frontier, and the stage was set for a backlash against the carnage. A movement to save America's beleaguered wildlife began to emerge, led by members of the eastern aristocracy. In 1887, for example, Theodore Roosevelt founded the Boone and Crockett Club as an association of sportsmen dedicated to preserving America's dwindling populations of big game. The

National Association of Audubon Societies, progenitor of today's National Audubon Society, was established in 1902 to protect the nation's birdlife. People, in short, were coming to the realization that North America was no longer a limitless frontier with endless supplies of wildlife. Both the animals and their habitats needed protection.

Here and there, small bands of bison were gathered up and protected on private ranches and preserves. (The motives of some of these ranchers were suspect, to say the least; they maintained small herds of bison in order to sell the rights to shoot them to unscrupulous, wealthy hunters.[17]) A few zoos began breeding the animals as well. Only one free-roaming herd remained in the United States, in Yellowstone National Park, and it continued to decline due to poachers until the federal government finally stepped up protection of the park and its wildlife. In Canada, a few wild bison survived in remote parts of Alberta and the Northwest Territories.

FROM THESE REMNANTS, something of a bison renaissance has come about over the course of the past century. Today, a half million or more bison occur in North America, and the future of the species seems secure. But numbers alone can be deceptive. According to a 2002 survey, approximately 96 percent of extant bison are being raised for commercial purposes.[18] Bison meat, which has markedly less cholesterol than beef, has become an increasingly popular choice of health-conscious Americans, and this, more than anything else, explains the growth in North America's bison population. Thus, the vast majority of today's bison are ranched bison, and the ranchers who manage the herds tend to select for docile, fat animals. They are, in effect, domesticating the American bison, turning it into a shaggier cow.

To make matters worse, a century ago ranchers crossbred a number of the surviving bison with cattle in an unsuccessful attempt to create a superior breed of livestock ("cattalo" and "beefalo").[19] Over the years, those hybrids were backcrossed with bison, with the result that most of today's bison herds are "contaminated" with cattle genes. Thus, if we ask how many genetically pure bison there are in the United States that are not being bred for commercial use, the answer is no more than a few thousand. And if we ask how many of those herds are free-roaming within their native range, our answer comes down to one: the population in Yellowstone National Park.[20]

And even that nominally protected population is under siege.[21] The problem this time is not hide hunters, but rather a small bacterium known as *Brucella abortus*. Native to the Old World, *B. abortus* was brought to North America via infected cattle imported from Europe. It spread to Yellowstone's bison around 1917 and now infects a significant fraction of the population. For the bison, it's not much of a problem since it produces little in the way of illness or disability. The symptoms in infected cattle are generally mild as well, but it does cause some infected cows to abort their fetuses and reduce their milk production, and in the low-profit-margin world of western ranching, those losses are intolerable.

Montana has worked hard to eliminate brucellosis (as the disease is called) from its livestock herds by testing each cow, destroying those that test positive for the bacterium, and by vaccinating calves. There is at least the theoretical possibility that wild bison could transmit brucellosis to cattle—it has happened in captivity when the two species were placed in close confinement—but the circumstances would certainly be unusual, such as a cow licking the afterbirth material from a bison within forty-eight hours of the bison giving birth. Should any of Montana's cattle contract the disease, the state would lose its brucellosis-free designation, a designation that allows ranchers to ship cattle outside the state without first quarantining them (a time-consuming and expensive step). That prospect has made state wildlife officials determined to keep Yellowstone's bison far away from Montana's cattle.

And there's the rub. While not migratory in the grand style of their plains ancestors, some of Yellowstone's bison nonetheless undertake an altitudinal migration during harsh winters, exiting the park and following established routes along the Yellowstone and Madison river valleys to lower-elevation sites where they have easier access to food. Some of the land they wander through is privately owned, and some of it is owned by the federal government but managed by agencies other than the Park Service, such as the U.S. Forest Service and Montana Fish, Wildlife and Parks. Ranchers are permitted to graze livestock in these areas, raising the possibility that bison could come into contact with cattle. The State of Montana has insisted that all bison wandering outside the park be quickly returned to the park or be killed, lest they infect the cattle. Because bison have a mind of their own and the brawn to exercise it, chasing them back into the park does not always work; all too often, the standoff is ended by a bullet. During the winter of 1996–97, for example, over a thousand Yellowstone bison (approximately a third of the park's population at the time) were killed by rangers and game wardens for the crime of migrating outside the park. Similar culling operations have taken place in subsequent years.

In response to public outrage over the killings, the State of Montana and the federal government developed an Interagency Bison Management Plan in December 2000.[22] Under this plan, bison that leave Yellowstone are chased back into the park through the use of helicopters, snowmobiles, all-terrain vehicles, and people on horseback. Those that refuse to return to the park are captured and tested for brucellosis. Individuals testing positive for brucellosis are killed, but up to one hundred brucellosis-free bison will be allowed to remain outside both the park's western and northern boundaries, provided the animals stay within the (arbitrary) boundaries of two bison management areas. Any bison that cannot be chased back into the park and that elude capture for testing will also be shot. Moreover, if the total population of bison within Yellowstone National Park tops three thousand, then all wandering individuals that cannot be chased back into the park may be shot without testing them for brucellosis.

The plan is meant to appease the ranchers while providing a fig leaf to hide the State of Montana's intolerance of bison outside the park. At best, it would allow two hundred sanitized bison to exit the park, provided the total population of bison in the park is below three thousand. To be fair to the State of Montana, the goal is not to eradicate Yellowstone's bison, but rather to eradicate their migratory behavior. Whether it's even possible to accomplish the one without the other is unknown.

Other solutions to the problem have been proposed and rejected. One is to round up all bison in the park, check them for brucellosis, and destroy those that test positive. Apart from the immense logistical difficulties of corralling, holding, and testing thousands of wild bison (in a national park no less!), the end result would surely be the destruction of a large fraction of the Yellowstone herd, inasmuch as 45–50 percent of the bison test positive for exposure to brucellosis.[23]

Others have suggested vaccinating the park's bison against brucellosis. In fact, a vaccine is currently being tested on a few of the park's bison, but its efficacy is low. Moreover, capturing and vaccinating bison inside the park on the scale necessary to eradicate the disease poses another logistical nightmare. In addition, bison are by no means the only wild animals that harbor the dreaded *B. abortus*. Elk are also carriers. Why ranchers and state officials are not equally alarmed about the presence of the disease in the region's elk population is difficult to fathom. Elk are apparently more fastidious in their birthing behavior than bison, with females consuming the placental tissues and fluids to avoid attracting predators.[24] This behavior may reduce the risk of transmitting brucellosis to livestock, but against this diminished risk of transmission per individual must be weighed the much greater population of elk—approximately one hundred thousand in the Yellowstone region. Indeed, there have been several documented cases of elk transmitting brucellosis to cattle in nearby Wyoming and Idaho.[25] Perhaps the main reason elk are treated more leniently by state officials is because of the importance of elk hunting to Montana's economy.

There are ways of dealing with brucellosis that are kinder to Yellowstone's bison. For example, the federal government could terminate or buy out the ranchers' grazing leases on the public lands surrounding the park as well as purchase key private lands where bison congregate in winter, in effect creating a "safe zone" where bison and cattle never overlap. This approach has little support among western legislators, who tend to be deferential to the cattle industry, if not obsequious. Thus, the killing continues, largely unnoticed by the public during mild winters when few bison leave the park, but hugely controversial during harsh winters, when lots of bison are killed. It would appear that even in Big Sky country, there is no longer room for a remnant of the American bison's grand migration.

Bison are by no means the only migratory mammals falling victim to human greed or indifference within the Yellowstone ecosystem. Yellowstone's bighorn sheep, elk, and pronghorn all engage in altitudinal migrations, abandoning the higher elevations during the winter and finding refuge and forage at lower levels. All three species have suffered greatly from the rapid development now under way across much of the West. Ecologist Joel Berger of the Wildlife Conservation Society recently estimated the number of pronghorn, elk, and bison migratory routes that have disappeared over the past century in the Yellowstone region.[26] He scoured the published literature, agency reports, and historical records such as trappers' journals to determine the locations of summer and winter ranges, as well as migratory routes. He then compared these historical data with the current situation, identifying winter and summer ranges that had been abandoned either because the ranges themselves had been developed or because the intervening migratory route had been blocked by development.

Berger estimates that within the twenty-seven-million-acre Yellowstone region, encompassing the park and surrounding federal, state, and private lands, no fewer than 58 percent of the elk migratory routes, 78 percent of the pronghorn routes, and 100 percent of the bison routes have been lost. He identifies several factors responsible for these losses. For bison, the major issue is intolerance. Many people do not want bison to exit the park, in part due to the brucellosis issue and in part because bison determined to get somewhere can plow through fences and corrals (a costly nuisance to landowners who must repair the fences and round up any livestock that exited through the gaps).

In the case of elk, kindness rather than intolerance is a major threat to migration. A number of years ago, the State of Wyoming created a network of feeding grounds where wild elk are provided with food during the winter months. These feeding grounds were established as compensation for losses of winter range due to development. Moreover, by providing the animals with a reliable food supply in winter, the state hopes to boost elk numbers for hunting. Unfortunately, the feeding grounds have caused many elk to curtail their migratory behavior.

Finally, a rapidly growing human population in the Yellowstone region, along with the associated loss of habitat, poses a threat to the migratory routes of virtually all the large mammals, from mule deer to bison. Moreover, the recent zealous, almost fanatical, push to develop the region's oil and gas resources could result in the creation of thousands, even tens of thousands, of new wells and drilling pads over the course of the next decade or two, further disrupting migratory routes.

Berger and his colleagues highlight one migration in particular: the 340-mile round-trip trek some pronghorn annually make from their summer grounds in Grand Teton National Park to their win-

Pronghorn antelopes

ter grounds in Wyoming's upper Green River Basin. It is probably the longest migration currently undertaken by any terrestrial mammal in the coterminous United States.[27] The pronghorn making this journey must contend with four natural bottlenecks. These occur in places where the terrain forms a narrow corridor, such as a pass between towering cliffs or across a mountain range. This inherently challenging journey has become vastly more difficult in recent decades, thanks to the 105 fences, miles of highways, acres of housing developments, and numerous drill pads for fossil fuels that have been placed along the way. Today, only about two hundred to three hundred pronghorn continue to make the trek. Given plans to expedite oil and gas development in the southern portion of the route, combined with a growing human population in the north, it seems only a matter of time until the last pronghorn faces a fence too high or one too many drill pads, thus ending a migratory tradition that has endured for thousands of years.

Although elk, pronghorn, and bison are in no danger of disappearing completely from Yellowstone, the loss of so many migratory routes, especially the long ones, surely diminishes the splendor of the region. Moreover, the Greater Yellowstone Ecosystem is one of the least-disturbed regions of the country. If it can suffer so many losses of migratory populations, even greater losses must be happening elsewhere in the West, as new housing developments, roads, strip malls, and other potential obstructions spring up across the landscape.

Berger and colleagues have proposed some sort of federal protection for migration corridors, analogous to the way we currently designate (and protect) national scenic highways, historic trails, and wild and scenic rivers. From an ecological perspective, the proposal makes eminent sense inasmuch as the current patchwork of conservation laws and regulations seems ill-suited to the task of safeguarding complete migratory corridors. But with the focus of so many elected officials now on resource extraction and development rather than conservation, there seems little chance of such a law being enacted any time soon. The fallback—turning to the individual states and pressuring them to protect migratory corridors— ignores the prime reason for federal action in the first place: many migration routes cross state borders and cannot be protected adequately in piecemeal fashion.

IN CONTRAST TO THE SITUATION in the Northern Rockies, bison, pronghorn, and other migratory species may be poised to make something of a comeback in the Great Plains. If this happens, it will not be the result of some grand federal plan, but rather a combination of economic, ecological, and sociological factors.

Even in the best of times—and those have been few and far between—the Great Plains are a tough place to make a living. Here one finds, in the words of Rutgers University geographers Frank and Deborah Popper, "the nation's hottest summers and coldest winters, greatest temperature swings, worst hail . . . , fiercest droughts and blizzards, and therefore its shortest growing season."[28] Yet the federal government has been trying for almost a century and a half to put people on the plains, first by offering them free land via the Homestead Act of 1862, then by providing crop subsidies and technical assistance (especially following the Dust Bowl calamity of the 1930s), and finally by financing an expensive infrastructure of dams and other irrigation projects. Despite these efforts, the region seems trapped in a cycle of boom and bust, with the bust years growing more frequent. For many of the ranchers and farmers, a good income is impossible, an adequate income is barely possible, and financial ruin is a real possibility. Younger people, faced with the prospect of a lifetime of hard work and little to show for it, are abandoning the farms and rural towns and heading elsewhere. As a result, the Great Plains is one of the few regions of the country to have seen a net decrease in population over the past two decades.

Pondering these trends, Frank and Deborah Popper proposed a

radical solution to the region's declining economics and social conditions in 1987: a Buffalo Commons. The Great Plains, they argued, cannot sustain intensive agriculture and livestock ranching—not economically, not ecologically, and not socially. What this region can do is grow bison, lots of bison, along with pronghorn, elk, sharp-tailed grouse, prairie dogs, and myriad other species adapted to living in this harsh land. The federal government can either stand back and witness the eventual depopulation of the Great Plains, the Poppers argued, or it can intervene to prevent this region from becoming "an utter wasteland, an American Empty Quarter."[29] They recommended that the federal government step in and buy back the land, then tear down the fences, remove the cattle, and restore the native wildlife.

Farmers and ranchers, not to mention governors and congressional representatives from the affected states, were openly hostile to the idea when it was first proposed. It seemed beyond the pale that two academics (from New Jersey, no less) would purport to tell these people that a way of life they had known for generations was unsustainable and unwise. But time appears to be proving the academics right, at least with respect to the broad outlines of their vision. As more farms and ranches face bankruptcy and as younger generations continue to desert the rural towns, some residents of the Great Plains are reconsidering their once-fierce opposition to the Buffalo Commons.

One such person is Mike Hayden. As governor of Kansas, he disputed the notion that rural communities were living on borrowed time. "Tell the Poppers that America's Great Plains do not equal the Sahara," he declared when the Poppers' article came out. Nearly two decades later (and no longer in office), his attitude has changed. "To stay the course is essentially a dead-end road," he now admits. "The Poppers were right."[30]

Up to a point, they *are* right. Something like a Buffalo Commons may be building in the Great Plains, but it is not the result of any bold action on the part of the federal government, which continues to dole out subsidies to traditional agricultural interests. Nor is it happening on quite the scale the Poppers envisioned. Instead, the initiative is being taken by nonprofit organizations like the American Prairie Foundation and the World Wildlife Fund. In less than four years, these two organizations quietly raised about ten million dollars, which the American Prairie Foundation used to purchase and lease 58,500 acres of ranchland in Phillips County, Montana.[31] The groups' goals are eventually to acquire several hundred thousand acres from willing sellers and to restore bison and other plains animals to the landscape. By teaming up with the nearby 1.1-million-acre Charles M. Russell National Wildlife Refuge, the organizations hope to create a large enough reserve to reestablish something close to a free-ranging, genetically pure bison population. It won't be a migratory population of the sort settlers encountered nearly two centuries ago. But the animals will surely move around over the course of the year in search of new forage, and in so doing they may create their own small-scale migrations.

In October 2005, sixteen healthy, brucellosis-free bison were captured in South Dakota's Wind Cave National Park and brought to the American Prairie Foundation's reserve in Phillips County. On November 17, before a delegation of reporters and conservationists, the gates of the acclimation corral were opened, and several minutes later the first brave bison ventured forth, quickly followed by the others.[32] The following spring, five calves were born, an auspicious start to the restoration effort. Meanwhile, the foundation is working to secure additional bison to increase the herd. Sometime soon, I intend to visit that reserve, preferably in the spring. The longspurs should be in fine form at that time of year, catapulting themselves into the air and singing lustily. But what I really hope to see is something far less graceful: an itchy bison rolling around on the ground, kicking up a cloud of dust and leaving a new and enduring imprint on the land.

In the Water ~

CHAPTER 5

Lost at Sea

I T ISN'T DIFFICULT TO imagine ways to block the migrations of terrestrial animals such as wildebeest or pronghorn. A sturdy fence will do; a new village or housing development is even better. But the notion that animal migrations in the oceans can also be blocked is more difficult to fathom. People are not building in the sea the way they are building on land. But slowly, surely, and with devastating effect, they are turning the migratory routes of many marine animals into deadly obstacle courses. Moreover, many ocean migrants share with terrestrial migrants the habit of congregating at key breeding, wintering, or stopover sites, which makes them all too easy to hunt. Consequently, for whales, sea turtles, and numerous other migratory species, the "freedom of the seas" is becoming more and more of an illusion as the ocean becomes an increasingly inhospitable environment for migration.

Few migratory animals face a more dangerous gauntlet than the North Atlantic right whale, now one of the rarest animals on earth. In the summer of 2004, marine biologist Chris Slay invited three of us to join him in a search for right whales. Our starting point was the town of Lubec, Maine, close to the Bay of Fundy.

Chris met us at the town dock in the early morning, towing a white Zodiac. To my landlubber's eyes, the Zodiac seemed ridiculously small for venturing into the bay, but Chris assured me it would do the job nicely. He maneuvered it into the water while I donned the mandatory survival suit—a bright orange flotation outfit that made me look like a fluorescent version of the Michelin Man. Uncomfortable as it seemed, the suit would keep me warm during the ride out into the bay and, more importantly, keep me alive for a few precious minutes in the unlikely event I ended up in the water.

As we crossed the Grand Manan Channel and headed into the bay, we spotted a few greater shearwaters and Wilson's storm-petrels, pelagic birds that rarely approach land except to breed. Both species nest on sub-Antarctic islands and migrate north to escape the austral winter (which corresponds to the temperate summer). Storm-petrels are little birds, no bigger than swallows, and as I watched them dance around the waves in search of tiny crustaceans and other morsels, I found it hard to believe that each one had traveled thousands of miles across the sea to reach the Bay of Fundy. Between sightings of shearwaters and storm-petrels, we also spotted a few puffins resting on the water. The burly, football-sized birds frantically pattered along the surface of the water before taking wing and veering out of the path of the Zodiac.

Chris had been in radio contact with two other whale research vessels on the bay that morning, and the early reports were none too encouraging. Right whales were few and far between. Finally, though, word came from one of the other boats, the New England Aquarium's *Nereid*, that it had just encountered a group of whales. We headed to the spot. When we were about a mile away from the *Nereid*, Chris stopped the Zodiac and we scanned the sea. Way off in the distance, we could see the occasional spout of a right whale, but around our boat the sea was as smooth as glass and equally life-less. Needless to say, I wanted to race right up to the other boat to share in its good fortune, but Chris kept us at a discrete distance.

Patience is not my strong suit—it never has been—but Chris knew what he was doing. After a half hour or so of waiting, an enormous black tail surfaced briefly less than a quarter mile from our Zodiac, followed shortly thereafter by a noticeably smaller tail. A cow and a calf were heading our way. Chris maneuvered the Zodiac near where he thought the animals might surface, and moments later, only a couple dozen yards off the bow, they did. The calf was staying near the surface of the water while its mother was diving for food. We soon spotted yet another cow-and-calf pair, then some more adults, and we quickly realized we were in the midst of a feeding aggregation of North Atlantic right whales. Chris guessed that somewhere between twenty and thirty individuals were foraging around us, perhaps 10 percent of the world's popula-tion of this species.[1]

We watched in awe as the whales lolled around near the sur-face, occasionally lifting a barn-door-sized flipper out of the water as though to give us a half-hearted wave, sometimes rolling over on their backs, and sometimes lifting their heads out of the water, revealing the deeply curved arc of their enormous jawlines and the peculiar callosities on top of their heads. These callosities consist of patches of callous skin covered by thousands upon thousands of brownish whale lice, ectoparasites that are probably unique to North Atlantic right whales and therefore equally endangered.

When a big, black tail slowly flipped out of the water and then sank back in, we knew an adult whale was deep-diving for copepods—little marine crustaceans that can be incredibly abundant at this time of year. During such a dive, a whale can easily hold its breath under water for fifteen to twenty minutes.

The most striking feature of the right whales (apart from their length) was their girth. Simply put, these animals were fat—exceptionally wide and round around the middle. According to Chris, Japanese whalers used to assume roughly 1 ton of weight per foot of length for all the great whales they hunted, with the exception of right whales. For right whales, they assumed 1.5 to 2 tons per foot, so laden with blubber is this species. A healthy right whale contains so much blubber, in fact, that upon being harpooned it floats rather than sinks. This characteristic, combined with its propensity to stay relatively close to shore, made it the "right" whale to hunt, thereby explaining its peculiar name.

At one point, a calf's curiosity got the better of it, and it came over to inspect us. Three or four times it stuck its head out of the water, each time venturing a little closer to the Zodiac. Several minutes went by without any sign of it, and I wondered whether the calf had grown bored with us (surely we were a rather uninteresting sight from its perspective, just a white object floating in the water, emitting unintelligible sounds). But Chris was not so sure it was gone. "It might be checking us out from below," he remarked, noting some peculiar, square areas of calm water, each the size of a dinner table, around us. These were water prints created by the whale's flukes. Suddenly, the young whale materialized alongside the boat, just a few feet underwater. It turned and swam directly under the Zodiac. "Stay inside the boat!" Chris cautioned. (Where else would I go, I wondered.) The calf circled around us a few more times before heading off to rejoin its mother. Gradually, the rest of the whales dispersed, leaving that particular stretch of the sea to the storm-petrels and puffins. Chris lowered the outboard motor

Migration of the North Atlantic right whale

into the water, started the engine, and pointed the Zodiac toward shore.

We had caught the North Atlantic right whales near the end point of their annual migration. During the late summer and fall, they congregate in the Bay of Fundy and adjacent waters off the coast of Nova Scotia. Here a combination of nutrient-rich waters and long days generates immense numbers of phytoplankton; the phytoplankton, in turn, sustain immense numbers of copepods, which are the favorite prey of the right whales. The whales simply open their cavernous mouths and swim through the swarms of copepods. Inside the whale's mouth, plates of baleen arrayed like rows of thick hair act as enormous sieves, capturing the copepods while allowing the water to pass through.

As fall gives way to winter, however, the feast ends. Waning sunlight and colder temperatures cause the copepods to go dormant and sink to the bottom of the sea, prompting the whales to disperse. The pregnant females head to warmer but less productive waters along the coast of Georgia and northern Florida, where they give birth to and nurse their young. Some (but not all) of the males and nonpregnant females migrate there as well. Where the others go remains a mystery, despite centuries of searching on the part of whalers, followed by scientists. They may disperse along the coastline or they may move farther out to sea. When summer returns, however, most of the whales will once again head to the rich waters of the Bay of Fundy.[2]

The North Atlantic right whale's migration thus appears to be driven by the same opportunism that causes birds to migrate to Maine or Nova Scotia: the northern latitudes provide abundant food, but for only a brief period of time. Unlike migratory birds, however, the right whales give birth on their winter grounds, probably because the waters of the Bay of Fundy become too cold too quickly for the babies. Therefore, the pregnant whales must store up enough blubber during their feeding binge in the north to sustain not only themselves but also their nursing calves during the lean winter months in the south.

THAT ANY NORTH ATLANTIC right whales are alive today to make this journey seems nothing short of miraculous, given how close this species has come to extinction. By all accounts, the North Atlantic right whale is living on borrowed time. At one time, the whales occurred on both sides of the Atlantic. While one population migrates between the Bay of Fundy and the southeastern United States, a second population once spent the summer in the waters off Scandinavia and wintered off the coast of southern Europe and northwestern Africa. Basque whalers pursued this European population at least as far back as the eleventh century.3 From October to March, lookouts on the northern coast of present-day Spain and France watched for the spouts of whales and alerted the boatmen, who headed out in pursuit of the animals. As their boats and whaling equipment improved, Basque whalers headed farther offshore into the Bay of Biscay. Whaling in the Bay of Biscay peaked around the mid- to late seventeenth century and declined quickly thereafter as the whales were depleted; by the 1720s, right whales were too few and far between in the eastern North Atlantic to be worth pursuing. Today, that population may be extinct.

Much of the blame for the swift demise of the eastern North Atlantic population can be attributed to the hunting methods used by the early whalers: they routinely searched for mother-calf pairs and harpooned the calf first, knowing that the mother would come to its aid. In this way the whalers could kill both animals. Such a tactic may have made sense in the short run, but it ensured the eventual extirpation of the whales.

As early as the fourteenth century, Basque whalers were scouring the English Channel and the waters off southern Ireland for right whales. By the mid-sixteenth century, they were hunting whales in Newfoundland, Labrador, the Gulf of St. Lawrence, and the Strait of Belle Isle, although their primary target was apparently bowhead whales, not right whales. Whaling off the American coast began in the 1630s, focused on North Atlantic right whales, and by the mid-1700s right whales were already well on their way to extinction.

At its nadir, the total population of North Atlantic right whales on both sides of the ocean probably numbered well under one hundred individuals.[4] Yet it took until the beginning of the twentieth century for most people to recognize that the species was on the cusp of extinction and in need of protection. The 1931 Convention for the Regulation of Whaling included a provision protecting right whales.[5] Moreover, the North Atlantic right whale was among the first group of species to be protected under the U.S. Endangered Species Act when that law was passed in 1973.

The question that troubles conservationists today is why, after decades of protection, the North Atlantic right whale has shown little sign of recovery—so little, in fact, that the official recovery plan for the species, prepared by the National Marine Fisheries Service, refuses to specify what number of whales would constitute a secure population because achieving any such target would be "too distant and hypothetical." Even the intermediate step in the recovery process—downgrading the species from endangered to threatened—is deemed decades away.[6]

The North Atlantic right whale's failure to rebound appears to be rooted in a mixture of biology and economics. They are slow-growing, low-reproducing animals. A female does not give birth to her first calf until she is, on average, nine years old. Gestation may take a year or longer, and weaning an additional eight to seventeen months.[7] The interval between births ranges from two to seven years and appears to have lengthened in recent years. Thus, the average lifetime reproductive output of a female North Atlantic right whale may be as low as 5.25 calves.[8] Given this low rate of reproduction, the population is limited in its ability to recover quickly. But even so, after more than a half century of protection, one would expect the right whales to be doing better (as, indeed, are many of the other species of whales that were once persecuted but are now protected). Instead, the most recent population models suggest the population is actually declining.

Here is where economics kicks in, for the North Atlantic right whale has the misfortune to migrate along one of the busiest coastal transportation routes in the world, and it spends its summer in a part of the western North Atlantic where fishing is a major industry. In the words of biologists Scott Kraus and Rosalind Rolland, it is now an urban whale.[9] Collisions with ships and fatal entanglements in fishing gear have replaced harpoons as the major threats to its survival. Of 50 deaths of North Atlantic right whales reported to wildlife authorities between 1986 and 2005, at least 19 were due to ship collisions and another 6 were due to entanglement.[10] Twenty-five deaths due to collisions and entanglement over a ten-year period may not seem like much, but for a species so rare, the loss of any individuals is cause for concern. Moreover, these numbers represent only the deaths we know about, which are presumably just a fraction of the actual number of right whales that perished during this time.

The agencies charged with protecting North Atlantic right whales have taken a number of steps to reduce mortality, usually in response to pressure from environmental organizations.[11] During key times of the year, for example, planes are sent out regularly to spot right whales, and the locations of all sightings are relayed to the U.S. Coast Guard and Navy, which broadcast them to ships operating in the area. The ships, however, are not required to reduce their speed or maintain a lookout for the animals, except while in the calving grounds. More significantly, some shipping lanes in Canada and in the Stellwagen bank off Massachusetts have been moved to reduce conflicts with right whales. But, by and large, there has been no serious effort to reroute commercial traffic away from the migratory route of the whales.[12]

Entanglement is an especially serious problem with respect to fisheries that use fixed gear, such as a string of lobster pots attached to a buoy or a gill net weighted down with sinkers. In the United States, state and federal wildlife officials have imposed seasonal restrictions on the use of such gear to protect the whales; they have also sponsored research to develop new types of fishing gear that are less likely to entangle whales, and they have created a network of first responders who attempt to disentangle any whales that are found ensnared in fishing gear. Thus far, the Canadian government has not imposed similar restrictions on fishing gear used within its borders.[13]

Not surprisingly, these half-hearted steps have failed to halt the decline of the North Atlantic right whale. During a sixteen-month period in 2004–5, eight deaths of right whales were reported to authorities, an unprecedented number of mortalities for such a short time period. At least three were killed by ships and one by fishing gear. This sudden loss of whales recently prompted a team of scientists to label the North Atlantic right whale a species "in crisis."[14] Unless ships are rerouted away from the whales' migratory pathways and unless the amount of fixed fishing gear in the water column is eliminated or severely reduced, the scientists declared, the North Atlantic right whale may be doomed. On the other hand, such bold steps would be certain to arouse the ire (and political clout) of the commercial shippers and the fishing industry (which undoubtedly explains the U.S. and Canadian governments' reluctance to take action).

Having spent a day in the company of right whales, I don't pretend to have a dispassionate view of their fate. That these giant yet innocuous animals have survived a thousand years of persecution by people astounds me. To lose them now, not because we must harvest the very last ones for our own survival or because they pose some threat to human well-being, but simply because they get in our way as they migrate, strikes me as unconscionable.

THE NORTH ATLANTIC RIGHT WHALE stands apart from most other whales in terms of its rarity and vulnerability, but not in terms of its migratory behavior. Most of the great whales, including bowhead, southern right, Pacific right, blue, fin, sei, Bryde's, minke, humpack, gray, and sperm whales, have one or more migratory populations, and these populations follow the same basic pattern.[15] Migratory whales living in the Northern Hemisphere spend their summer months in the high latitudes, feasting on the enormous stocks of zooplankton and fish that occur there seasonally. As the weather gets colder and the sea ice begins to form, they retreat to lower latitudes in the seas around North and Central America, Eurasia, or Africa. Whales living in the Southern Hemisphere typically spend the austral summer (i.e., December, January, and February) in the high southern latitudes, sometimes venturing as far south as the edge of the Antarctic pack ice. Here, too, the long days and fertile seas produce an abundance of krill and other marine life that sustains not only whales, but also vast numbers of seals, sea lions, and seabirds. As summer gives way to winter, these whales retreat to lower latitudes around South America, South Africa, and the Pacific Islands. In short, it's the mirror image of what happens in the north.

Most of the great whales have separate populations in the Northern and Southern Hemispheres, as well as along opposite sides of the major oceans. These separate populations typically follow different migratory pathways. What no whale attempts to do, apparently, is to transverse the globe, much as the Arctic tern does, spending the austral summer (the Northern Hemisphere's winter) near Antarctica and the temperate summer in the High Arctic. While doing so might enable the animal to enjoy each pole's seasonal smorgasbord, the journey in between would be impossibly difficult. It would take a huge amount of time, and the whales would probably overheat while passing through the warm waters of the tropics.

How whales navigate between their summer and winter quarters remains a mystery, largely due to the difficulties associated with studying them. As noted in chapter 1, a major obstacle to studying migratory birds is their small size; only recently have engineers developed transmitters small enough to attach to a thrush or sandpiper. On the other hand, scientists have been bringing birds into the lab for decades in order to place them in strange magnetic fields, expose them to different star patterns beneath the dome of a planetarium, or otherwise alter their sensory world, all for the purpose of understanding what navigational cues they use during migration. In the case of whales, the opposite problem applies. There is no practical limit to the size of the transmitter one can attach to a whale, but it is impossible to bring the giant animals into the lab for experiments.

Some scientists have hypothesized that whales let the topography of the coastline or seafloor guide them, perhaps by using the changes in water depth or flow that result from undersea ridges as underwater "landmarks" in much the same way migrating birds use coastlines or mountain ranges. Others have suggested that whales use changes in the earth's magnetic field as a compass-and-map system to guide them, again analogous to what some birds are able to do.[16] The small amounts of magnetic iron oxide that have been found inside the brains of whales lend support to the magnetic-field hypothesis, but without direct experimental evidence, one cannot be certain. And, of course, it is possible, even likely, that whales rely on more than one navigational system for their long-distance migrations.

During the heyday of international whaling, from the second half of the nineteenth century to the early years of the twentieth, when factory-sized whaling ships prowled the seas from the Arctic to the Antarctic, the migratory behavior of the great whales has-tened their demise in two ways. First, to the extent that any species followed predictable routes or congregated in key feeding areas, it became an easy target. Second, the very fact that the whales were migratory gave people a false sense of security with respect to their numbers. When a particular species declined in a given area (coin-cidentally the area in which hunting was heaviest), the whalers simply assumed the animals had moved somewhere else. Asked if he was concerned about declines in humpback whale populations in the Southern Ocean, one whaling captain replied, "I am absolutely of the opinion that no danger threatens, as the hump-back has his undisturbed haunts in the ocean; we have many proofs that he is a migratory animal going from ocean to ocean."[17]

Willfully blind to the future, pursuing their prey on the high seas far outside the jurisdiction of any government (and often with the tacit support of their home-port governments), commercial whalers managed to deplete stock after stock of whales. They start-ed with the species that lived closest to shore—right whales and gray whales—and when those were gone, they turned to the ones that lived farther offshore or were slightly harder to catch, like the humpback whales. Then, as their boats grew larger and the easily caught whales grew scarcer, the whalers ventured farther and far-ther out to sea, in search of the giant blue and fin whales, until these too were depleted; then the small but numerous minke whales became the target.

As early as 1906, the governor of the Falkland Islands felt com-pelled to introduce legislation to restrict whaling operations under his jurisdiction in response to evidence of overexploitation. But given that the whaling industry was moving farther and farther into international waters, something other than a piecemeal, nation-by-nation approach was needed. This realization on the part of several nations led to the 1931 Convention for the Regulation of Whaling (the agreement that, among other things, protected right whales) and its successor, the International Convention for the Regulation of Whaling, which came into force in 1948. The latter agreement created the International Whaling Commission (IWC), which remains the principal international body charged with regulating the harvest of whales.[18]

Throughout the 1950s, it became increasingly clear to scientists that whale stocks were declining as whaling continued. But scientific evidence of declines was not enough to prompt the IWC to make the necessary cuts in harvest levels. Year after year, pressure from the whaling industry overrode science. Moreover, as was discovered only after the collapse of the Soviet Union, catch data submitted by the Soviets to the IWC in the 1960s and 1970s routinely underreported the number of whales that were killed. The situation, in short, was worse than anyone imagined.

In 1982, the IWC finally imposed a moratorium on all commercial whaling, beginning in the 1985–86 season. Norway lodged an immediate objection to the moratorium. Under terms of the 1948 treaty, a nation lodging an immediate objection is allowed to continue a legal commercial harvest, which Norway has done. Iceland and Japan also objected, although neither nation filed a formal objection. However, both countries have continued to hunt whales under a provision of the treaty that allows for limited harvests for scientific purposes. Between 1986 and 2004, the two nations harvested over ten thousand whales under the banner of science. The actual scientific value of those whaling operations has been called into question many times, especially since the meat of the harvested animals is sold in restaurants and food stores (with the approval of the IWC, which insists that the flesh of whales killed for scientific purposes not go to waste).[19]

Eager to increase their allowable harvest and aware of how flimsy the case for "scientific" whaling really is, proponents of whaling have adopted a new line of attack: by killing more whales, they argue, we can increase the amount of seafood available for humans, since a reduced whale population cannot consume as much marine life. Given our current overexploitation of the oceans, such an argument is bound to appeal to many countries with declining fisheries. The reasoning, however, flounders in the face of studies showing that most of the food consumed by large whales consists of species that fisheries do not target. Moreover, the areas where the great whales are most abundant are not the hot spots for commercial fishing. In terms of diet and preferred fishing grounds, the overlap between humans and whales is minimal.[20]

That people are even debating whether to resume commercial whaling is testimony to the benefits of the restrictions imposed by the IWC and other regulatory bodies. A number of whale populations have rebounded since commercial whaling was stopped, including humpback whales in the western North Atlantic, gray whales in the eastern North Pacific, and bowhead whales in the Bering, Chukchi, and Beaufort seas.[21] For other species, accurate trend data are unavailable. Population estimates for some of the more numerous whales, especially the minke, suggest these animals are common enough to sustain a limited commercial harvest, but whether the whaling industry can be trusted to abide by the rules is another matter. By analyzing the DNA from whale meat sold in Japanese markets, biologists have demonstrated that even under the banner of "scientific" whaling, the wrong species are often killed and sold, including fully protected ones like humpback, blue, and fin whales.[22] Moreover, science cannot address the moral revulsion that millions of people feel toward any type of whaling, and this sentiment, more than statistics or population models, surely underlies much of the current opposition to a resumption of commercial whaling.

OF ALL THE GREAT WHALES, none has staged a more dramatic comeback than the gray whale. Centuries ago, there were populations of gray whales in the North Atlantic, eastern Pacific, and western Pacific. The North Atlantic gray whales disappeared some-time in the late seventeenth or early eighteenth century, victims of overexploitation. The western Pacific population summers in the waters around northern Japan and the Korean Peninsula and win-ters off the southern coast of South Korea. Never very numerous, this population was hunted extensively and now teeters on the brink of extinction. The eastern Pacific population, which migrates between Alaska and Baja California, slid perilously close to extinc-tion as well in the early decades of the twentieth century. However, this population has subsequently rebounded to the point at which gray whales are as abundant today along the Pacific Coast of North America as they were three centuries ago. Their annual trek from Alaska to Baja and back may be the longest migration of any mammal. The migrating whales attract thousands of eager whale-watchers every year; the whale-watchers, in turn, sustain a growing ecotourism business in Mexico, the United States, and Canada. For students of animal migration, the gray whale's tale rep-resents that rarest of events: a bona fide conservation victory in contrast to the declines of so many other migratory species.

For the vast majority of gray whales in the eastern Pacific, life begins inside one of a half-dozen sheltered lagoons scattered along the coast of Baja California.[23] Here, in the warm, shallow waters, females give birth in the winter months, following a gestation of a year or more. Unmated females winter in the lagoons as well, where they are joined by boisterous crowds of adult males eager to impregnate them. Thus, the lagoons function as both nurseries and singles bars. By late February or March, some of the single males and females are ready to begin their northward migration; the oth-ers will follow them in the weeks to come. The last to leave are usually the pregnant females and nursing mothers, some of whom linger around Baja until well into May.

The whales hug the coast as they head north, traveling 4,000 to 5,000 miles until they reach the Bering Sea. Some even cross the

Bering Strait and head around the north slope of Alaska into the Beaufort Sea. Progress is slow but steady: One radio-tagged gray whale traveled from Baja to Unimak Pass, Alaska—a distance of 4,142 miles—in ninety-four days.[24] It averaged 44 miles per day over the course of three months. Once the whales reach the Bering Sea, they gorge on small crustaceans that they vacuum out of the sediments on the seafloor. Gray whales put on tons of blubber during this time, which they will rely on to sustain themselves for much of the rest of the year, since they rarely feed during migration or while in Baja. Come October, with daylight disappearing and temperatures dropping, the whales leave the Arctic and head south—pregnant females first, followed by adult males and non-pregnant females, with juveniles last. The vanguard will reach the Baja lagoons in late December or early January; the others will arrive during January and February. (The staggered nature of the southbound migration, coupled with the early departure of some individuals from Baja, means that the last of the southbound whales may cross paths with the first of the northbound whales.)

Until the middle of the nineteenth century, gray whales were not a priority target for whalers operating along the Pacific Coast of the United States and Mexico, largely because they yielded less oil than other species. However, the discovery of the winter lagoons quickly changed that assessment. Big numbers of gray whales could be harvested quickly and efficiently, and the whalers jumped at the opportunity.

The first of the lagoons to be discovered was Bahia Magdalena in the winter of 1845–46. From then until 1873–74, approximately 2,200 gray whales were taken there by whalers. The next to be found was Laguna Ojo de Liebre, in the winter of 1857–58. The discoverer, Captain Charles Melville Scammon, returned to San Francisco in the summer of 1858 with 700 barrels of oil, the product of about 20 gray whales.[25] Captains of other whaling vessels must have been dumbstruck when they learned of Scammon's cargo, for he had obtained in eight months' time as much oil as other ships might obtain in four years. Needless to say, the next winter many other whaling vessels raced to Laguna Ojo de Liebre, eager to partake of the bounty. Over the next four years, some 22,250 barrels of oil, the product of more than 600 gray whales, were removed from Laguna Ojo de Liebre until very few whales were left in the lagoon. The same thing happened to the other lagoons as well. In the 1850s, the gray whale population in the eastern Pacific may have numbered more than 20,000; two decades later, fewer than 2,000 remained. Norwegian whalers, operating out of large factory ships, continued to take gray whales in Mexican waters throughout the first two decades of the twentieth century. By 1930, some scientists estimated that no more than a few dozen gray whales remained in the entire eastern Pacific.

The U.S. government added the gray whale to its endangered species list in 1970. The following year, Mexican president Luis Echeverría Álvarez signed legislation designating Laguna Ojo de Liebre as the world's first whale sanctuary. Another important whale site, Laguna San Ignacio, was added to the protected roster in 1972. Environmentalists within Mexico and around the world continued to push for more protection of the lagoons and the surrounding desert, and in November 1988 the Mexican government protected a seven-million-acre area around Laguna Ojo de Liebre and Laguna San Ignacio as the Vizcaíno Desert Biosphere Reserve. The aim of the reserve was, and is, to conserve the area's wildlife and natural values by ensuring that only compatible human activities are allowed within its borders.[26] The key breeding areas, it appears, are now reasonably safe.

Other threats to the gray whales are more diffuse and thus more difficult to address.[27] Collisions with ships and entanglement in fishing gear have claimed the lives of a number of gray whales, although neither problem at present appears to be harming the overall population. A more significant threat may be noise disturbance.

Playback experiments have shown that the sounds of drilling ships and drilling platforms can frighten away migrating gray whales. There is also concern that the acoustical pulses generated by the air guns and water guns used to conduct seismic surveys do not merely frighten whales, but actually harm them. Given growing pressure for offshore oil and gas development, coupled with more ship traffic up and down the Pacific Coast, the gray whales face an increasingly noisy and dangerous migration in the years ahead.

That said, they clearly have benefited from the protection they have received thus far. Current estimates place the eastern Pacific gray whale population at well over twenty thousand individuals, more or less their number prior to exploitation. In 1994, the U.S. government removed the gray whale from the endangered species list, declaring it recovered.[28]

The rise in gray whale numbers has been accompanied by an equally sharp rise in the number of tourists who flock to see them.[29] Today, there are whale-watching enterprises all along the Pacific Coast, from Alaska to Mexico, based wholly or partially on gray whales. Indeed, because they undertake such a long migration so close to shore, the gray whales enable many coastal communities to benefit economically from whale-watching. A study published in 2001 tallied 65 whale-watching operators in California, 10 in Oregon, and 26 in Washington, serving more than 878,000 customers per year. In addition, every year hundreds of thousands of people enjoy the migrating gray whales simply by observing them from strategic points along the shoreline. Total annual revenues stemming from ship- and land-based whale-watching in those three states were estimated at more than $83 million in 1998. Although other types of whales, most notably humpbacks and orcas, generate some of this income, the gray whale is the prime draw in southern California and Oregon.

Mexico, too, is developing a thriving whale-watching industry. In the 1970s and 1980s, local fishermen began taking a few tourists out into the gray whale lagoons aboard small boats called pangas. Most of these guides operated as subcontractors to U.S. tour operators, and relatively little of the money generated by the tourism benefited the local communities. In the mid-1990s, however, more and more Mexican operators became involved in the whale-watching business, as international interest in gray whales increased and as the tourism infrastructure in Baja improved. By 1998, there were close to 20 operators along the coast of Baja serving over 35,000 whale-watchers per year. These tourists, in turn, generated roughly $5.4 million per year in revenues, much of which went directly into the local communities.[30]

Thanks to tourism, people up and down the Pacific Coast now have a direct economic stake in the welfare of gray whales. Recognizing that this link may be critical to the long-term protection of the whales' habitat, especially in Mexico, several conservation organizations have worked to expand local involvement in ecotourism. One of the more successful initiatives was launched by Rare, a nonprofit conservation organization that seeks to protect biodiversity by enabling local communities to benefit from preservation efforts.

In the late 1990s, Rare began a multiyear campaign to enable Mexicans living near the lagoons to benefit from the growing interest in whale-watching. The group first identified a few individuals living around Bahia Magdalena and Laguna San Ignacio who it thought had the potential to become successful nature guides. Rare then put them through an immersion course in English (the primary language of ecotourists visiting Baja, so English-speaking guides command a premium), as well as intensive classes on the natural history of gray whales and other wildlife in and around the lagoons. Rare then assisted the graduates in building relationships with the major tour companies bringing tourists to Baja. When these people obtained employment as nature guides, interest in Rare's program grew; thus far, fifty-seven people have received training. As a result, most of the thirty thousand tourists who visit the lagoons nowadays are guided by Rare-trained personnel. For the nature guides, Rare's involvement has meant new livelihoods and greater prosperity. For the whales, according to Rare's president, Brett Jenks, "this means more conscientious boat drivers, better compliance with zoning and use regulations, better managed protected areas, public support for the Vizcaino Biosphere Reserve . . . and a direct link between conservation priorities and community self-interest."[31]

As gray whales became more valuable alive than dead, something remarkable happened to the whales themselves: they stopped fearing people. During the heyday of whaling in the nineteenth century, the gray whale was known as the devil fish, based largely on incidents in which gray whales attacked whaling ships. (That most of those incidents occurred after the ship had harpooned either the mother or her calf seems not to have been a mitigating factor in the minds of the whalers.) Then, in the early 1970s, several decades after the cessation of commercial whaling, Mexican fishermen were astonished when a few gray whales began approaching their boats and allowing themselves to be petted. The whales seemed to enjoy the contact, even to seek it out. The behavior spread among the whales, and today quite a few regularly approach boats to interact with people. To the best of my knowledge, no one has come up with a compelling explanation for this change in behavior on the part of the whales. I suspect they simply enjoy getting their heads scratched.

There is reason to be cautiously optimistic about the future of the great whales. Populations of several species have rebounded as a result of protection (although some, like the North Atlantic right whale, are still in great danger of extinction). In addition, whale-watching is growing in popularity around the world. By one estimate, the number of whale-watchers grew by more than 12 percent per year between 1991 and 1998, and the number of countries and overseas territories with whale-watching enterprises jumped from thirty-one to eighty-seven during this same period.[32] With so many people now smitten by whales, it's hard to imagine a return to the days of widespread commercial hunting (although there are certainly a few nations that would support such a move).

I AM FAR LESS SANGUINE about the future of another group of oceanic travelers, the sea turtles. All seven of the world's sea turtle species—from the 100-pound Kemp's ridley to the 1,000-pound leatherback—are endangered. Like whales, they are magnificently adapted to life at sea. Over 110 million years ago, they exchanged the stumpy legs of their land-dwelling ancestors for elegant flippers that enable them to swim for thousands of miles across the oceans. And by slowing their heartbeat to conserve oxygen and by redirecting their blood flow to critical regions of the brain, heart, and central nervous system, they can stay under water for long periods of time, up to several hours in the case of some species.

However, sea turtles never evolved the ability to incubate their eggs under water; instead, they rely on a warm, sandy beach to accomplish that essential task. So, unlike whales, female sea turtles must come ashore periodically to reproduce. Out of this imperative has evolved one of the oddest migrations in the animal kingdom: every two to four years (depending on the species), a female sea turtle returns to the land—usually to the very beach where she was born—to lay her eggs. Although a female sea turtle probably spends less than one-tenth of 1 percent of her life ashore, this tenuous, relictual connection to the land greatly magnifies the dangers of migration, exposing the animal to threats ranging from egg-eating raccoons to beach-devouring developers. Sea turtles, in short, face many of the same hazards confronting whales, plus an additional set of hazards derived from their habit of nesting on land.

To be on a beach when a sea turtle comes ashore to lay its eggs is to witness something primordial and awe inspiring. On a warm

Loggerhead turtle

June evening in 2004, Conservation International scientist Will Turner and I traveled to the Archie Carr National Wildlife Refuge, a protected stretch of beach along Florida's Atlantic coast, to see this phenomenon. Our host was Lew Ehrhart, a biology professor at the University of Central Florida. For more than three decades, Lew and his students have been studying green, loggerhead, and leatherback sea turtles that come ashore each year to nest along this stretch of the Florida coast.

Once the sun had set, Lew dispatched teams of scouts to work their way up and down the beach in search of nesting sea turtles. We didn't have to wait long: within fifteen minutes, the research station's radio crackled with word that a female loggerhead had been spotted on the beach.

Lew hustled us down the beach, warning us not to use our flashlights or talk loudly. When they first come ashore, sea turtles can be rather skittish; bright lights or strange sounds will cause them to turn around and head back into the ocean. It took my eyes a while to adjust to the darkness, but when they finally did I could make out a distant dark boulder crawling out of the surf. As we got closer, the rock turned into a female loggerhead. We watched silently as she dragged herself—all four hundred pounds—up the beach in the direction of the sand dunes. That she wasn't having an especially easy time of it seemed obvious. The sea turtle huffed and wheezed as she dug her flippers into the soft sand and hauled herself toward the dunes, pausing from time to time to catch her breath.

When she reached the base of the dunes, she stopped. Some cue had penetrated her reptilian brain and told her that this was the spot to lay her eggs. Whatever that cue was, it triggered a deeply ingrained, stereotypical program for digging the actual nest. She brushed the sand toward the center of the nest with one of her rear flippers, then scooped it up and flung it out with the other. Over and over she repeated this behavior, until she had constructed a small pit behind her. After a short pause, the egg-laying commenced, with the eggs, each about the size and shape of a Ping-Pong ball, dropping into the hole in groups of one to four. I attempted to tally them as they exited the turtle but quickly lost count. (I subsequently learned that Florida's loggerheads on average lay 112 eggs per clutch and produce slightly less than four clutches per breeding cycle.)

Once the turtle began laying her eggs, Lew allowed us to relax our guard. At this stage neither noise nor lights would dissuade a sea turtle from completing its business and, indeed, our turtle seemed to be in some sort of trance, oblivious to everything around her save for the task of laying her eggs. When the last egg had been deposited, she methodically filled the pit with sand and slowly, laboriously made her way back to the surf. Abandoning any pretext of scientific objectivity, I imagined how relieved she must have felt as those first waves washed against her, marking the end of her exhausting journey ashore.

I LEFT THE Archie Carr National Wildlife Refuge later that evening. Had I been able to return in eight or nine weeks, I might have witnessed another remarkable sight: dozens upon dozens of tiny loggerhead turtles, each no bigger than a half-dollar and weighing less than an ounce, crawling out of the sand at night and scrambling to the sea. This mad dash through a potential gauntlet of predators, including raccoons, herons, and gulls, marks the beginning of a sea turtle's long journey.

Upon leaving the nest, the hatchlings appear to have an innate urge to head toward the brightest area on the horizon. (This can be demonstrated by placing them in little arenas with a movable spotlight; they will scurry to whatever side has the light.) Because the hatchlings do not emerge until dark, this simple behavioral rule is enough to guide them to the ocean, where starlight reflected off the water creates a brighter horizon. Once in the water, the turtles head into the waves. Doing so takes them away from the shoreline with its many predators and out toward the open ocean. Once in the ocean, the young loggerheads may linger in places where the surface waters converge to form downwellings, hiding out and feeding in the rafts of seaweed that typically collect in such places. Or they may ride the Gulf Stream across the Atlantic Ocean to the Azores. Regardless of where they go, the little turtles will spend anywhere from six to twelve years in their pelagic nurseries, eating and growing and eluding predators. (Such youthful wanderings are hardly restricted to loggerheads in the Atlantic Ocean; juvenile loggerheads captured off the coast of Baja California have been traced to hatching sites in Japan.[33] Other species of sea turtles are suspected of making similarly grand journeys after hatching as well.)

Eventually, the young loggerheads find their way back across the ocean to coastal inlets, sounds, bays, and estuaries. Here they may stay for another decade or two until they are somewhere between twenty-five and thirty-five years old, at which point they are finally ready to reproduce. After mating, each female will make her way back to her natal beach—to a place she has not seen since she was one day old—to lay her eggs.

How the turtles accomplish this remarkable feat of homing remains something of a mystery, despite decades of research. In a recent experiment, biologists captured young green turtles in the ocean off Melbourne Beach, Florida, and tethered them in a circular arena that was filled with water and surrounded by coils that could be used to manipulate the ambient magnetic field.[34] When the turtles were exposed to a magnetic field equivalent to the one existing roughly two hundred miles to the north of Melbourne Beach, they swam in a southerly direction, as though they were trying to get back home. When they were exposed to a field equivalent to the one existing two hundred miles south of Melbourne Beach, they swam in a northerly direction. The results demonstrate that sea turtles can use the earth's magnetic field to orient themselves on a north–south axis (a talent they share with birds and perhaps monarch butterflies and whales). Whether they can also use it to determine their precise location along an east–west axis (a necessary step in order to have an accurate mental map of where they are) is less certain. It's quite possible they can, but they may use other cues as well, such as topographical features like coastlines or even the smell of particular beaches and inlets.

After laying their eggs, adult female loggerheads return to the ocean. They may forage in the Gulf of Mexico, the Caribbean, or along the Atlantic Coast of North America, sometimes venturing as far north as Newfoundland. Those that head north in the spring apparently turn around and return south in the fall as the water temperature drops. These loggerheads are therefore engaging in a pattern of metamigrations: yearly movements up and down the coast in response to seasonal changes in water temperature and, after that epic first migratory return to the natal beach, a trek back to the nesting beach every two to four years to lay eggs.

Green and hawksbill turtles more or less follow the same pattern as the loggerheads: after egg laying, they swim to feeding grounds relatively close to shore, where they stay until they are ready to nest again in two to four years. Leatherback turtles, by contrast, appear to wander the oceans after nesting, traveling thousands of miles until the urge to reproduce once again draws them back to their natal beaches. Adult leatherbacks tagged in French Guiana, for example, have been recovered in Newfoundland; individuals tagged in Papua New Guinea have been found in Monterey Bay, California.[35] If one were to tally all the miles traversed by a leatherback during the years it spends at sea, it might well emerge as the greatest traveler in the animal kingdom.

The annual wanderings that adult loggerheads make up and down coastlines are easily explained as examples of migratory opportunism: in such cases, one presumes the turtles are taking advantage of a flush of food available in the northern latitudes during the warm months and retreating south as the weather (and water) turns colder. It's similar to what warblers or whales do. On the other hand, the journeys that loggerheads and all other sea turtles make to and from their nesting beaches are qualitatively different from what other migratory animals do. The turtles are coming ashore solely to lay their eggs, and it's hard to see what, if anything, is special about the particular beaches they choose. Why the turtles come back to these same beaches year after year and generation after generation remains a puzzle.

The sea turtles' connection to the land, as noted earlier, means they face virtually every threat a marine creature must endure plus an additional set of dangers that come with being a big, slow-moving, clumsy animal on land. For untold thousands of years, people have been raiding sea turtle nests or capturing the adult animals when they come ashore, but the turtles nonetheless persisted. As the age of sail dawned, the intensity of the hunting began to mount, however. It may not be much of an exaggeration to say that sea turtles literally fueled the exploration of much of the world from the sixteenth to the nineteenth century. Sailors walked the beaches at night, flipping over the adult females. The next day, crews loaded the hapless animals onto small boats, took them to their ships, and filled the holds with them. Since sea turtles can survive for weeks or even months without food, the sailors were assured of a source of fresh meat during their long voyages.[36] In the short term, this harvest made sense. The turtles were abundant, easy to capture, and even easier to store. But in the long term, the relentless consumption of adults and eggs sent many turtle populations into a tailspin, forcing the sailors to search widely for new nesting colonies to exploit.

Dwindling populations of sea turtles prompted local and national governments to enact laws limiting the harvest of adult turtles and their eggs, with some statutes dating back to the early seventeenth century.[37] However, adult sea turtles and their eggs continue to be harvested in many developing nations, often because the governments fail to enforce the laws that protect sea turtles and their nests.[38]

Yet even if the adults and their eggs were well protected, sea turtles would still be in trouble due to the expansion of fishing around the world. As fishing fleets have grown in size and technological sophistication, they have taken an increasing toll on the world's sea turtles—not intentionally, but accidentally, as the animals become trapped in nets or hooked on lines.

Among the more deadly tools of modern fishing are gill nets. Designed so that a fish can poke its head through the mesh but is ensnared by its gills if it tries to back out, gill nets are superb at catching fish. But they are also superb at catching sea turtles, which blunder into them and drown unless the nets are checked frequently. In the 1980s, fishers in the Pacific Ocean employed gill nets that were hundreds of miles long and hundreds of feet deep, resulting in the drowning of thousands of sea turtles. As a result of an international agreement, gill nets of this magnitude are now banned on the open sea. But smaller ones continue to be used in nearshore waters, where they remain a threat to sea turtles.

Longlines are another deadly problem. To anyone who still thinks of fishing in terms of a monofilament line with a single hook at the end, today's longline vessels will come as something of a shock: multiple lines are set, each extending for miles and covered with hundreds or thousands of baited hooks. In 2000, according to one scientist, fishers from forty nations set at least 1.4 billion hooks on longlines that averaged forty miles each in length.[39] These lines have been implicated in the deaths each year of hundreds of thousands of seabirds (which grab at the bait as the hooks hit the water) and sea turtles (which either become entangled in the line or take the bait when the hooks are underwater). There are ways to discourage the birds, largely by sinking the lines faster or by placing a line of streamer ribbons just above the spot where the hooks are being placed into the water, but figuring out how to discourage the sea turtles has proved more difficult.

Shrimp trawling is another major source of mortality for sea turtles. In many places around the world, both shrimp and sea turtles frequent the same coastal waters, with the result that shrimp nets annually ensnare and drown thousands of sea turtles. Fortunately, this particular problem lends itself to a technological solution: a mesh or wire structure known as a turtle excluder device (or TED).[40] When attached to a trawl net, it allows the turtles to escape. First developed in the United States in 1980, TEDs were hardly welcomed by the shrimp-fishing industry. To the contrary, fishers spent well over a decade fighting any attempts by the federal government to force them to use TEDs when fishing in certain waters inhabited by sea turtles, despite a clear requirement under the Endangered Species Act that they do so.

Their refusal to use TEDs was based largely on fears that the devices would reduce their catch, either by allowing shrimp to escape or by causing the nets to become clogged with seaweed. In fact, studies revealed that TEDs can boost the shrimp catch by reducing the amount of other sea life—bycatch in the jargon of fishing—that is accidentally caught in the nets. Nonetheless, U.S. fishers fought the use of TEDs until all their legal (and, on occasion, illegal) options were exhausted. In developing countries, fishers have been similarly reluctant to adopt TEDs, often citing the same reason given by their U.S. counterparts, namely the fear that TEDs will reduce their catch.[41] Additional hurdles in such countries include the cost of acquiring the devices and the lack of training in their proper use.

Sea turtles face plenty of problems on land as well. The beaches where they nest are coming under increasing pressure from developers, as anyone who has traveled along Florida's Atlantic coast can attest. Here the issue is not just the conversion of once-prime nesting sites to condos, resorts, and boardwalks; a host of more subtle problems are associated with even modest development near the nesting beaches. Outdoor lighting is a good example. As noted previously, turtle hatchlings head for the brightest spot on the horizon when they emerge from the sand in the predawn darkness, a strategy that for millions of years has led them to the surf. But if that brightest spot is now the porch light of a beach house, the hatchlings are likely to wander in that direction rather than toward the sea, usually with fatal results. (Fortunately, homeowners can limit the harm caused by artificial lighting by using low-pressure sodium lights, which appear not to disorient the turtles as much as do incandescent bulbs.) Garbage is another problem. People produce trash, and trash attracts raccoons and other scavengers. When not raiding garbage cans, these scavengers are likely to be raiding turtle nests on the beach.[42] More people living near the coast means more trash, and more trash means more scavengers, to the detriment of sea turtles.

And, like all marine organisms, sea turtles are vulnerable to pollution. Leatherback turtles, for example, apparently mistake floating plastic bags and deflated balloons for their much-loved jellyfish, an understandable mistake for a creature that evolved over a hundred million years before the invention of plastic. Unfortunately, their lack of discrimination often leads to fatal stomach obstructions. According to the U.S. Fish and Wildlife Service, nearly half the adult leatherbacks sampled worldwide had plastic in their stomachs.

There may also be a link between marine pollution and disease outbreaks in sea turtles. In recent years, scientists have detected a high incidence of cutaneous fibropapillomatosis in certain populations of sea turtles. This disease causes the growth of large, disfiguring tumors. If the tumors grow externally, they can interfere with a turtle's ability to swim or eat; if they grow internally, they can damage internal organs such as the lungs, stomach, and kidneys. Because the disease is especially common among turtles living near large human populations with poor sewage treatment facilities, some biologists suspect it is triggered by viruses, bacteria, or other pollutants associated with people.

Given the myriad threats sea turtles face, it hardly comes as a surprise that all of them are in danger of extinction. What we tend to forget in our efforts to save today's dwindling populations is just how far they have fallen, how numerous sea turtles once were, and how important they may have been in the ecology of marine ecosystems. By examining historical records and doing some back-of-the-envelope arithmetic, ecologists Loren McClenachan, Jeremy Jackson, and Marah Newman have estimated that the pre-Columbian populations of green turtles and hawksbill turtles in the Caribbean Sea alone may have numbered over 90 million and 11 million, respectively.[43] Other estimates have placed the population of green turtles in the Caribbean as high as 660 million.[44] At those numbers, the living weight or biomass of green turtles in the Caribbean would have approached or exceeded the biomass of bison on the Great Plains. This comparison is particularly apt when one considers that green turtles, like bison, are grazers: they feed on sea grasses. The green turtles kept the sea grasses short and productive by continually cropping them, and the turtles also redistributed nutrients via their urine and feces. It is not too far-fetched to imagine that green turtles were once as important eco-logically in the Caribbean Sea as bison were in the Great Plains.

Building on this work, sea turtle researcher James Spotila has estimated that more than a billion sea turtles roamed the world's oceans prior to human persecution, including hundreds of millions of greens, tens of millions of loggerheads, and several million leatherbacks.[45] And today? The Pacific Ocean may contain as few as 2,300 adult female leatherbacks; the population of green turtles

in the Caribbean is less than 300,000; Caribbean hawksbills probably number in the low tens of thousands.[46] We struggle to protect
tiny remnants of once vast aggregations.

THERE IS LITTLE PROSPECT of restoring the world's sea turtles
to anything approaching their pre-Columbian grandeur. But we can
aspire to increase today's tiny populations. Doing so is not rocket
science, but it is a significant economic and political challenge.
Nesting beaches must be protected against further development
(often to the dismay of developers), and the adult turtles and their
eggs must be protected from poachers. Both steps, although politi-
cally vexing, are straightforward and can often be accomplished by
individual nations acting on their own.

Addressing the mortality caused by trawlers and longlines, on
the other hand, entails regulating the activities of vast numbers of
fishers around the world, including those who operate in international waters. The mandated use of TEDs on shrimp trawlers is an
obvious first step, but few nations have laws as strong as the U.S.
Endangered Species Act to compel such a change in fishing practices. Moreover, in developing countries, fishers often lack the
money to purchase TEDs, and even if they had the financial where-
withal, there are no institutions ready and able to teach them how
to use the devices properly. Since 1989, the United States has
attempted to spur the use of TEDs in other countries by prohibit-
ing the importation of shrimp captured in ways deemed harmful to
sea turtles. Countries that require and enforce the use of TEDs are
allowed to sell shrimp in the United States.

With respect to longlines, there is currently no proven way to
make them safe for sea turtles, although some promising developments with respect to the design of the hooks may eventually
reduce sea turtle mortality. Until technology delivers a bona fide
solution, the best hope for the turtles lies in the willingness of

governments to prohibit the use of longlines within their territorial waters and in the willingness of consumers to boycott seafood captured in turtle-hostile ways.

Given that many sea turtles do not reach sexual maturity until they are over twenty years old, one would not expect their populations to rebound quickly, even if they are well protected. Patience should be the buzzword of sea turtle conservation. And yet, scientists have been amazed at how quickly turtle populations around the globe have responded to even modest conservation measures. Over the past thirty years, for example, the number of green turtles nesting on French Frigate Shoals in the Hawaiian archipelago has increased fourfold, corresponding to the turtles' protection under the Endangered Species Act.[47] Similar increases have occurred over roughly the same time period in Tortuguero, Costa Rica, home to the largest green turtle nesting aggregation in the Atlantic basin. This increase followed restrictions on egg and turtle collecting that were imposed by the Costa Rican government, as well as a decision by adjacent Nicaragua to ban the commercial harvest and export of green turtles.[48]

Even the Kemp's ridley, the rarest of the world's sea turtles, has shown signs of recovery in recent years.[49] Virtually all the remaining Kemp's ridleys come ashore to nest at Rancho Nuevo in Tamaulipas, Mexico, located approximately two hundred miles south of Brownsville, Texas. When first discovered in the late 1940s, this nesting aggregation contained upwards of 40,000 turtles. By the mid-1960s, only about 2,000 females were coming ashore to lay their eggs each year; by the early 1980s, fewer than 300 were doing so. The decline was traced to two problems: poaching of adult turtles and their eggs in Mexico, and accidental capture and drowning in shrimp nets off the coasts of Mexico and the United States. In the late 1970s, the U.S. and Mexican governments created a joint recovery team to save the Kemp's ridleys. The nests at Rancho Nuevo received added protection, and fishers were required to use TEDs in waters inhabited by the turtles.

Conservationists were troubled that almost the entire population nested on only one beach. A single catastrophe there could send the species spiraling to extinction. Thus, in 1978 the U.S. and Mexican governments launched an ambitious effort to create an additional nesting colony in south Texas by "head-starting" young Kemp's ridleys. Eggs were taken from Rancho Nuevo, brought to Texas, and incubated indoors. The newly hatched turtles were then allowed to scramble down to the surf at Padre Island, Texas, the hope being that they would imprint on this particular beach. The hatchlings were then rounded up and kept in captivity for an additional nine to twelve months before being returned to the wild. The additional time in captivity was intended to allow them to grow big enough to avoid predation upon release.[50]

Thousands of young Kemp's ridleys were released under the head-start program. As the years went by and no adults showed up on Padre Island, a growing number of scientists began to worry that the whole project was a failure. Some felt the hatchlings were not being given enough time to imprint on the beach; others believed the time in captivity was somehow rendering them unfit for life in the wild. Yet, over the past few years, a growing number of Kemp's ridleys have been showing up to nest on South Padre Island—over one hundred nests in 2006 alone. It may well be that the only thing missing from the head-start program has been patience. Kemp's ridleys apparently require more years to mature than scientists had predicted. Meanwhile, back at Ranch Nuevo, the added protection appears to be paying off. The number of Kemp's ridleys returning to nest has increased roughly tenfold from the low point two decades ago.

Stories such as these justify some measure of hope that we can stop the downward slide of the sea turtles and even rebuild some populations. It's worth emphasizing, however, that almost all our trend data are based on counts of adult females coming ashore to nest. This is arguably the most important segment of the sea turtle population to monitor, but it is just one segment. Scientists rarely know how juveniles or adult males are faring. Also, when they count adult females on nesting beaches, what they really are counting are the results of conservation measures taken many years ago. As ecologist David Ehrenfeld eloquently noted, "Looking at green turtle population data is like looking at the light from a star 25 light years away. It appears to be shining now, but, in fact, you are looking at history, and there is no way of telling whether, during the past 25 light years, that star has increased in brightness or perhaps has gone out altogether."[51]

Moreover, all our efforts to save sea turtles may prove futile if we fail to address a much bigger problem looming, quite literally, on the horizon: global climate change. Global warming could affect sea turtles in at least two ways. First, according to the latest climate models, if we do nothing to curb global warming, sea level is expected to rise anywhere from seven and a twenty-four inches over the next century, largely as a result of thermal expansion of the oceans and glacial melting.[52] Such a rise would be enough to submerge many of the beaches where today's sea turtles nest. In undeveloped areas, it may be possible for the beaches (and the turtles) to simply move inland as sea level rises. But in places such as Florida, where residential and commercial developments abut the beaches, there will be nowhere for the beaches or the turtles to go.

Second, sea turtles share with some other reptiles a peculiar trait: the incubation temperature of the eggs determines the sex of the offspring. For most species of sea turtles, the warmer the temperature of the sand, the higher the proportion of female offspring. In the case of green turtles, for example, an incubation temperature of 82 degrees Fahrenheit will produce male hatchlings, whereas at

88 degrees the clutch will consist of females. Intermediate temperatures yield a mixture of males and females.[53] The temperature/gender rules for the other species of sea turtles are similar, although the details may vary by a few degrees in either direction. Even the heat generated by the developing eggs themselves is sufficient to alter sex ratios within the nest. The warmer eggs in the center of the clutch tend to develop into females; those at the periphery are more likely to become males. Although all this may seem like a strange way to determine gender, the system has presumably worked well for sea turtles for over one hundred million years. But that was before humans began tinkering with the earth's climate.

Sea turtles are truly innocuous creatures, incapable of inflicting harm on humans and content to do what they have done for millions of years, which is to wander the oceans, munch on jellyfish, sea grass, or other delicacies, and, from time to time, make a pilgrimage ashore to lay their eggs. I refuse to believe that we are either so needy or greedy that we cannot share a bit of the land and sea with them. And I hope we are not too short-sighted to realize that altering the earth's climate could be as harmful to us as it will be to them.

CHAPTER 6

Against the Flow

IN THE DAYS FOLLOWING my encounter with the right whales, I spent a lot of time doing very little—very little, that is, other than staring out over the waters of Cobscook Bay from the deck of my friends' home in Lubec, Maine. Silvery-gray terns flew back and forth, carrying little fish they had caught to some distant islet where their insatiable chicks were waiting to be fed. Simultaneously, crews of men buzzed across the bay each morning in motorboats, tending to a dozen or so cages anchored in the water. Each cage contained tens of thousands of Atlantic salmon. These fish farmers were working hard to provide fresh salmon to a growing human population with a growing appetite for seafood. It eventually dawned on me: both they and the terns were in the same business.

Salmon farming has become a major enterprise in Maine, with

statewide production climbing from roughly 44,000 pounds in 1984 to over 13 million pounds in 2003. And quiet Cobscook Bay is now the epicenter of that industry. Counting all the fish in cages in the bay and elsewhere along the coast, there are undoubtedly more Atlantic salmon in Maine today than at any time in history. But to see a truly wild Atlantic salmon in the state—a fish that hatched from an egg laid by a wild female in the gravel of a Maine river—one has to be extraordinarily lucky. In 2006, the Maine Atlantic Salmon Commission counted fewer than 1,200 wild individuals heading upstream to spawn in the state's rivers. The Penobscot River alone accounted for over 1,000 of these fish, leaving only about 100 wild salmon in all the rest of Maine's rivers.[1]

From a global perspective, of course, the Atlantic salmon is in no danger of extinction. In addition to the millions raised each year in fish farms around the world, there are wild populations in Canada, Greenland, Iceland, Europe, and Russia. But within the United States, Maine is the wild Atlantic salmon's last redoubt.[2] And what is happening to salmon in Maine is symptomatic of what is happening to salmon in many other places around the world, where once-great spawning runs have slowed to a trickle, and government agencies and private individuals find themselves struggling to protect and restore wild fish populations in the face of competing demands for food, fiber, energy, and commerce.

The vulnerability of salmon stems in large part from their migratory behavior. Their habit of spawning upstream creates a natural bottleneck that makes it easy for people to catch them, in much the same way sea turtles become vulnerable to overexploitation when they come ashore to lay their eggs. Also, that salmon inhabit two distinct worlds—freshwater streams and the ocean— doubles their exposure to natural and human-caused changes. In other words, changes to either environment can be deadly to them. Not surprisingly, protecting these fish poses enormous economic, political, and social challenges. Unfortunately, each state or nation tends to repeat the same mistakes as it grapples with the challenge of salmon conservation, namely a futile search for a painless technological fix, followed by unproductive arguments over who or what is really to blame for the problem, until finally the various factions realize there is no shortage of blame to go around and no quick or easy way to restore salmon.

At one time, Atlantic salmon spawned in virtually every major river in the northeastern United States and eastern Canada, from Long Island Sound to the Ungava Peninsula in northern Quebec. Historical accounts suggest the spawning runs in some of these rivers were enormous. Prior to 1700, for example, few merchants in New England bothered to sell salmon because the fish were simply too abundant to command a decent price (often less than a penny per pound). Farther north, the territorial governor Nicolas Denys complained that Atlantic salmon were so numerous in New Brunswick's Miramichi River that the sound of them leaping out of the water and falling back in as they migrated upstream kept people awake at night.3 By the early 1800s, however, wild Atlantic salmon had disappeared from Long Island Sound, followed by southern New England in the mid-1800s, due to overharvest, dams, development, and pollution. Maine became the last holdout of the species in the United States, but even here salmon continued to decline as the same old threats caught up with them.

To fully understand the reasons behind the disappearance of Atlantic salmon, as well as the challenges associated with restoring them, it is helpful to know something about their complex life history. Atlantic salmon begin their lives as little eggs laid in the summer or fall and buried beneath the gravel of streambeds. The eggs develop slowly in the cold winter waters, eventually hatching in March or April. The newly hatched fish, called alevins, remain hidden in the gravel, living off their yolk sacs for a month or so. They eventually emerge from the gravel and begin feeding on plankton and tiny invertebrates, at which point they are called fry.

Fry quickly develop dark marks along their sides, denoting their entry into the so-called parr stage. Salmon parr remain in the rivers for one to three years, usually in riffles with moderate to fast flows and plenty of gravel and rocks for shelter.

When the parr have grown to four or five inches in length, they undergo an amazing transformation that prepares them for life in saltwater. Their body shape becomes more streamlined; the dark marks along the sides disappear, and the fish become silvery in color; they switch from facing upstream while swimming to facing downstream. The most important changes, however, are physiological. The osmotic functioning of the gills and kidneys must be reversed, lest water escape the fish's body when it enters the ocean, causing concentrations of salts and other minerals within the cells to reach dangerously high levels. (While in freshwater, a fish faces the opposite dilemma; it must work to maintain a sufficiently high concentration of salts and other minerals within its body.) After the makeover is complete, the young salmon (now called smolts) begin their journey to the sea, the first leg of their migration.

Exactly where they go and what they do when they reach the sea remains one of the bigger mysteries in salmon ecology, since smolts are too small to track via transmitters and satellites. Maine's smolts appear to spend their first winter in the Labrador Sea, south of Greenland, feasting on fish and other aquatic life in the productive marine environment. A few will return to Maine the following spring to spawn. Most, however, will spend another full year in the Labrador Sea before returning to breed; a few will even spend two more years in the ocean before heading back. When the salmon are ready to spawn, virtually all of them will return to the very streams where they were born.

The mechanisms by which they accomplish this remarkable feat of navigation are poorly known, especially with respect to how they find their way from the Labrador Sea back to the Maine coast.[4] However, biologists do know that as the fish get closer to shore, nearer to the mouths of the rivers, their sense of smell takes over. Each river, by virtue of its intrinsic chemical composition and the runoff from its shores, has a distinctive smell that salmon are able to perceive, and the fish apparently use that scent to locate their natal rivers. (Indeed, scientists have tricked salmon into swimming up the wrong river by imprinting them on an artificial smell in their natal river and then adding that smell to another river nearby.)

Thus, following a scent they last smelled one to three years previously, the salmon head back up the very rivers where they were born. Once in the river, they may pause for a few days or a few weeks before making the final push into the smaller streams where they actually spawn. The females search for suitable places to lay their eggs, usually riffles with a gravel substrate. Each female then digs a nest with her tail and deposits some seven thousand to eight thousand eggs. The eggs, in turn, are fertilized by sperm released into the water by nearby males. In Maine, most of the spawning takes place between mid-October and mid-November.

Other species of salmon living in the Pacific Ocean typically die right after spawning, having given that last full measure of devotion to procreation. In the case of Atlantic salmon, however, anywhere from 5 to 25 percent of first-time spawners (depending on the location) go back to sea and then return to spawn a second time the following year.[5] Fewer than one in a hundred will survive to spawn a third time. Thus, for most Atlantic salmon, one shot at reproduction is all they get, even if they have not exhausted themselves to quite the same degree as their Pacific cousins.[6]

ACCUSTOMED AS WE ARE TO THINKING about migration as opportunism, it's easy to guess why the young salmon head out to sea. The marine environment is more productive than the streams where they were born. Thus, by spending one to three years in the ocean, a salmon smolt can grow much larger and invest more energy in reproduction than would otherwise be possible. Less clear is why, over the course of time, they have not completely abandoned the streams in favor of a purely marine life. One possibility is that salmon and their relatives evolved in a freshwater environment, and it is simply too difficult (from an evolutionary perspective) for eggs and fry to develop the physiological traits necessary for marine survival, whereas it is possible for adults to make the switch.[7] Another possibility is that by sequestering their vulnerable egg and parr stages in freshwater streams, salmon are able to escape many of the predators that lurk in estuaries, bays, and coves. If so, then a mixed strategy—giving birth in the streams but growing up in the sea—makes sense.

As we have seen with respect to other species, a life-history strategy that works brilliantly in a preindustrial world can be a major liability in a world dominated by people. In the case of Maine's salmon, consider first the changes that have occurred in and around the spawning streams. In southern and central Maine, most of the forests were cleared for farms. Soil washed off the farm fields and into nearby streams and rivers, smothering eggs and fry. The sediment pollution was also harmful to smolts. Moreover, as farming practices evolved, new problems arose. For example, farmers turned to insecticides, herbicides, and synthetic fertilizers to increase crop yields. These chemicals, in turn, found their way into the region's lakes and rivers, where they are toxic to young salmon. (Much of the land cleared for agriculture in the nineteenth century was subsequently abandoned, allowing the forests to regenerate. This regeneration has been beneficial for salmon.)

In the northern parts of the state, logging rather than farming was the predominant land use. Loggers high-graded the forests during the first half of the nineteenth century, removing the biggest and best trees for saw timber. Maine's timber industry experienced a slowdown in the mid-1800s as loggers turned their attention to the uncut forests of the Adirondacks and northern Great Lakes, but it experienced a major recovery in the 1880s with the development of wood-pulp paper. The advent of this new product meant that even small, low-value trees could be converted into profitable pulpwood, leading to record-high levels of logging in the first decade of the twentieth century.[8]

Unfortunately, good times for loggers meant bad times for salmon. Soil washed off clear-cuts and logging roads into streams, smothering salmon eggs and fry. The removal of tall, shady trees from the stream banks raised water temperatures to unhealthy levels for the surviving eggs and fry. And in order to send logs downriver more easily, timber companies cleared, bulldozed, and dynamited the channels to remove any boulders or fallen trees. Those boulders and natural tree falls, however, created deep pools and other habitat features that were important to salmon fry and parr. (The destructive practice of log drives continued well into the 1960s and did not stop completely until the 1970s.[9])

As the region's human population grew, so did the demand for energy, prompting the construction of numerous dams. Although some efforts were made to design dams that salmon could traverse, no such design works flawlessly, and many stretches of river suddenly became inaccessible to migratory fish. Moreover, even the addition of a fish ladder or other passage device cannot undo the harm the dam does to both the upstream and downstream portions of the river. The pools of slow-moving, warm water that form behind dams are typically unsuitable for salmon fry and parr, which prefer cooler, faster-moving streams. Below a dam, the flow pattern or hydrograph of the river is also disrupted, and this too can affect the temperature, chemistry, and other properties of the river in ways harmful to salmon. In sum, therefore, the freshwater component of the salmon's world was disrupted by three major types of habitat degradation: agriculture, forestry, and dams.

Adding to the Atlantic salmon's woes has been the release of numerous nonnative fish into Maine's lakes and rivers, most of which were deliberately introduced for the benefit of recreational fishers. These alien fish include potential predators of young salmon (e.g., chain pickerel, smallmouth bass, and largemouth bass) as well as competitors (e.g., brown trout, rainbow trout). Some, like the chain pickerel and smallmouth bass, were introduced well over a century ago; others are more recent additions. Although the State of Maine now tries to restrict releases of nonnative fish in rivers where wild Atlantic salmon occur, many of these introduced species are so well established and so widespread that there is little anyone can do about them. They constitute yet another type of habitat degradation.

The Atlantic salmon's connection to freshwater also makes the species vulnerable to another, subtler form of habitat degradation: acid rain.[10] Field and laboratory studies have demonstrated that salmon fry, parr, and smolts are very sensitive to increases in the acidity of the water in which they live. Smolts in particular are readily harmed by even a temporary increase in acidity, perhaps because they are undergoing that profound physiological transformation from a freshwater organism to a saltwater organism. Moreover, smolts in acidified streams exhibit behavioral changes: they tend to wander back and forth at the interface of freshwater and saltwater instead of heading directly to sea. This change in migratory behavior may expose them to greater predation pressures inasmuch as many predatory fish and birds dwell at this interface.

Although both the Canadian and U.S. governments have taken major steps to combat acid rain over the past three decades, the problem has by no means disappeared. The focus of most of these acid-rain reduction efforts has been on sulfur, and the results, measured in terms of reductions in the amount of polluting sulfur emitted by industry or vehicles, have been substantial. However, there is also a nitrogen component to the acid-rain problem, stemming from industrial activity and transportation, and it has proved far more difficult to control with existing technology. Thus, streams continue to acidify and salmon continue to die.

In addition to habitat destruction and degradation, Atlantic salmon have had to contend with overfishing. Their migratory behavior creates a natural bottleneck that people have been all too willing to exploit. In New England and Canada, for example, colonists regularly employed nets and fish-trapping weirs that spanned almost the entire width of rivers in order to catch salmon heading upstream.[11] Such tactics were highly successful in the short term, but disastrous over the long term. In a number of major rivers, salmon populations plummeted because most of the adults were picked off before they could spawn. By the time of the American Revolution, colonial legislators were introducing bills to prohibit river-spanning nets and other destructive fishing practices. But these early efforts to prevent overfishing in the rivers were, to some extent, undone by the growing armada of fishing boats that pursued the adult salmon at sea. Ocean fishing became an increasingly serious threat from the 1960s onward, when large trawlers equipped with state-of-the-art fishing technology began operating off the coast of Greenland.

Changes in populations of other migratory fish species have also had an effect on Atlantic salmon. Numerous other species, including alewives, blueback herring, American shad, sea lamprey, anadromous rainbow trout, Atlantic sturgeon, shortnose sturgeon, and American eel, also regularly migrate into and out of Maine's rivers. Some, like the salmon, spawn in freshwater but mature in the ocean; others, like the eels, grow up in freshwater but return to the ocean to spawn. Thus, Maine's rivers, like major rivers throughout the North Atlantic, are literally highways of fish coming and going in both directions. This whole community of fish may have been important to the long-term welfare of the salmon. Adult alewives, for example, migrate upstream to spawn at roughly the same time salmon smolt head downstream. Because alewives are about the same size as smolt but orders of magnitude more abundant, they likely serve as a prey buffer for the salmon. In other words, predation pressure on salmon by cormorants, otters, ospreys, and other animals is diluted by the presence of the abundant alewives. The alewives and other migratory fish also serve to transport nutrients from the productive sea to the less productive river, whether in the form of feces or their own decaying bodies after death. These nutrients, in turn, are essential for the growth and development of young salmon in the fry and parr stages.

Of course, the same dams, clear-cuts, farms, factories, and roads that have harmed salmon populations have harmed these other migratory species as well. The shortnose sturgeon, for example, was added to the federal endangered species list in 1967. The others, although in no danger of extinction, have nonetheless suffered major population declines, and these declines have only added to the salmon's woes.

HOW HAVE PEOPLE RESPONDED to the demise of wild Atlantic salmon in Maine and elsewhere? A report prepared by Maine's Commissioners of Fisheries as far back as 1869 recommended four steps to save the state's dwindling salmon numbers: (1) construct fishways over impassible dams; (2) curb overfishing; (3) protect water quality in the rivers where the fish spawn; and (4) in limited cases, restock rivers with hatchery-reared fish. Perhaps not surprisingly, the only one of these recommendations the state took to heart was the last, since it stepped on no one's toes.[12] By the early 1870s, fisheries managers were adding salmon parr and eggs to various Maine rivers in an effort to restore depleted populations. Meanwhile, the dam operators blamed the fishers for the decline in salmon, and the fishers blamed the dam operators, while the habitat itself was largely ignored.

Hatcheries (which rear and release young salmon) may seem like a logical solution to the salmon problem, but they are at best an imperfect and temporary fix. At worst, they can actually hasten the demise of wild salmon. The reasons are threefold. First, there may be important genetic differences among salmon populations spawning in different rivers. If each population is adapted to the unique conditions of the river where it spawns, then an infusion of hatchery-reared fish from other rivers can destroy the genetic integrity of those populations. Second, the very act of rearing fish in captivity selects for domesticity, which is undesirable in the wild. Third, a focus on hatcheries may cause people to ignore the underlying problems affecting the wild fish. If a hatchery can churn out millions of salmon, regulators and political leaders may feel less pressure to prosecute polluters, clamp down on timber companies and farmers, restrict development within watersheds, or take any of the other difficult steps that may be necessary to ensure the long-term health of the rivers and the fish within them.

The practice of stocking rivers with hatchery-reared fish continues to the present day, although the practice has become far more sophisticated than it was a century ago. Fisheries managers now strive to restock rivers with the spawn of locally captured fish so as to preserve the genetic integrity of the individual populations. Care is taken to minimize the effects of domesticity and inbreeding on the brood stock. Nonetheless, in 2004 a panel convened by the National Research Council to study the status of Atlantic salmon in Maine reported that "the available information is not sufficient to conclude whether hatcheries in Maine can actually help to rehabilitate salmon populations, whether they might even be harming them, or whether other factors are affecting salmon so strongly that they overwhelm any good that hatcheries might do."[13]

As it became clear that hatcheries alone could not rescue plummeting salmon populations, people began to consider other factors. A federal fishery plan enacted in 1987 prohibits the harvest of Atlantic salmon within the U.S. exclusive economic zone, which extends two hundred nautical miles off the U.S. coast. Then, in December 2000, the National Marine Fisheries Service and the U.S. Fish and Wildlife Service declared Maine's wild Atlantic salmon to be an endangered species. The decision reflected some of the scientific and political subtleties of conservation.

Maine is the last place in the United States where wild Atlantic salmon occur, and those fish are clearly in danger of extinction. Protection was therefore extended to wild Atlantic salmon from the Kennebec River northward to the mouth of the St. Croix River. However, the Fish and Wildlife Service and National Marine Fisheries Service excluded salmon in the Penobscot River, allegedly because that river has been heavily stocked over the years and therefore the salmon spawning within it may no longer represent a distinct wild gene pool. Yet the Penobscot has by far the largest remaining salmon population in Maine; the decision to exclude the river effectively denied federal protection to 90 percent of the state's (and the nation's) remaining wild Atlantic salmon. Some conservationists believe the real motivation behind this decision lay not in genetics but in the fact that the Penobscot River watershed drains a large portion of Maine. Including the Penobscot salmon would have greatly expanded the reach of the Endangered Species Act and engendered a strong backlash from the state's politicians, who feared the economic consequences of protecting the salmon and their habitat.

Indeed, when the salmon were added to the endangered species list, there were plenty of predictions of economic devastation. Alarmists warned that whole industries, from agriculture to logging to energy production, would be crushed by heavy-handed regulators. In fact, none of that has happened. The Endangered Species Act protects wild Atlantic salmon from intentional capture by either commercial or recreational fishers within the United States, although some fish continue to be caught unintentionally by both groups as they pursue other species. The activities of timber companies, blueberry growers, and farmers are now monitored with greater care, thereby reducing the risk to the fish from clearcuts, logging roads, improperly installed culverts, insecticides, and herbicides. The blueberry growers, in particular, have had to shift their irrigation practices away from drawing water out of salmon streams to using water from wells and reservoirs.[14] Equally important, adding the salmon to the endangered species list has resulted in an influx of federal money for salmon conservation efforts and has heightened cooperation among interest groups and state and federal authorities.

Even Maine's hydroelectric dams and other infrastructure are no longer quite the insurmountable obstacles they once seemed to be. A number have been removed or breached (not necessarily out of concern for fish; some were simply deemed uneconomical or unnecessary), the most notable example being Edwards Dam.[15] Built in 1837 near the mouth of the Kennebec River, the dam was a barrier to a number of migratory fish, including shad, striped bass, blueback herring, alewives, sturgeon, and salmon. Because the dam was no longer of any real value for hydropower, irrigation, or flood protection, the Federal Energy Regulatory Commission revoked its operating license in 1997, after years of lobbying by environmentalists. Two years later, it was demolished, thereby opening up an additional seventeen miles of the river to migratory fish. Striped bass, alewives, and other species have already begun to reclaim this stretch of the river, and, one hopes, it won't be long before Atlantic salmon do so as well.

Still other dams in Maine have been retrofitted with fishways to permit migration. Perhaps the most ambitious effort along these lines is one involving a coalition of conservation organizations and the Penobscot Indian Nation working together under the banner of the Penobscot River Restoration Project.[16] The project's goal is to purchase three dams that currently block salmon migration on the Penobscot River. Two of the dams will be dismantled; the third will be decommissioned and retrofitted with a fishway. If successful, these steps will provide salmon and other migratory fish with improved access to roughly five hundred miles of the Penobscot

River, while resulting in only a small reduction in power generation. Both state and federal agencies and the owner of the dam, the PPL Corporation, are cooperating in this effort. Restoration programs such as these are essential if salmon are to achieve any sort of real comeback. More than half of the spawning habitat originally available to salmon in Maine is no longer accessible due to dams and other man-made obstructions, and until those obstructions are modified or removed, the rivers in question can play no useful role in the recovery of Maine's salmon.[17]

Even with the Endangered Species Act, the U.S. government has relatively little ability to influence what happens to its salmon when they cross national borders or head into international waters. Fortunately, a number of other countries have taken an interest in the plight of the Atlantic salmon, not so much for the sake of the United States but rather to protect their own indigenous runs, which intermingle with U.S. fish when all are at sea. Canada, for example, has banned commercial fishing of Atlantic salmon in Newfoundland and Labrador, where many of Maine's fish reside during portions of their time at sea. In August 2002, Greenland suspended commercial fishing for salmon in its waters, where substantial numbers of U.S. and Canadian salmon spend part of their lives. However, Greenland did not end the harvest for domestic use, which recently has averaged slightly over twenty metric tons of salmon per year. How many of those fish are from Maine's tiny population is unknown. The U.S. Fish and Wildlife Service does not consider Greenland's internal fishery to be a threat to the survival of Maine's salmon, but it does view the harvest as a potential impediment to recovery of the species.[18]

Today, one of the biggest remaining threats to wild salmon in Maine and elsewhere is, oddly enough, farmed salmon, like the ones being raised in the pens I saw in Cobscook Bay. A combination of crowded conditions and frequent imports of new fish makes aquaculture pens veritable incubators of disease. The pathogens, in turn, can be transferred to wild salmon when wild fish swim too close to the pens or when captive fish escape. Among the more serious diseases is infectious salmon anemia, or ISA, a contagious and untreatable virus that damages a fish's kidneys and circulatory system. Mortality rates range from 3 to over 50 percent of infected individuals. Experiments have shown that ISA is transmitted via mucus, feces, and blood; hatcheries and fish farms therefore provide ideal conditions for an outbreak. Relatives of the Atlantic salmon, such as brown trout and rainbow trout (both introduced into Maine's rivers), can be asymptomatic carriers of ISA, providing additional pathways for the disease to spread.

An outbreak of ISA hit Cobscook Bay's salmon farms in 2000, striking first on the Canadian side and subsequently spreading to the U.S. part of the bay. By September 2001, half of the salmon pens in the bay were infected by the virus. In an effort to contain the epidemic, the Maine Department of Marine Resources ordered the destruction of one and a half million farmed salmon. Fish farmers were required to remove all salmon from the bay and to clean and disinfect their pens, boats, and other equipment. They also were prohibited from bringing any new fish into the bay for a period of sixty to ninety days. In addition, the Department of Marine Resources put in place new regulations that tightened husbandry standards, increased the monitoring and testing of fish farms for disease, and restricted the movement of boats and other equipment. By the spring of 2002, the state was ready to allow a resumption of salmon farming in Cobscook Bay. One year later, however, ISA was back, demonstrating just how difficult it is to avoid disease outbreaks when huge numbers of fish are maintained at high densities.

Nor is disease the only problem associated with salmon farming. Sea lice are frequent parasites of salmon when the fish are in the ocean (the lice quickly die when salmon enter freshwater). Under normal conditions, the density of sea lice per salmon is not high enough to cause the fish any serious harm, but in crowded pens the parasites can reach levels high enough to weaken or even kill the fish. Wild salmon, in turn, are vulnerable to infestation if they happen to migrate by the pens. (This is probably a more serious issue for young salmon heading out to sea than it is for adult salmon heading inland to spawn; the latter will lose their sea lice once they reach freshwater.) In addition, a study conducted in Norway suggests that sea lice can spread ISA from one salmon to another.

Also worrisome to some environmentalists is the frequency with which farmed salmon escape from pens. The biggest breakouts have been associated with storms that breach the pens or uproot them from their moorings. In December 2000, for example, a powerful nor'easter in Machias Bay, Maine, enabled more than one hundred thousand farmed salmon to escape into the wild—a number equal to roughly one hundred times Maine's entire wild salmon population.[19] The concern here is threefold. First, escaped fish can spread diseases to wild salmon; second, they may compete with wild salmon for food or spawning sites; and third, they may interbreed with the wild stock, thereby contaminating the wild gene pool.

Two other issues associated with salmon farming have less to do with their effect on wild salmon than with their effect on the overall health of the oceans. The first is pollution. A large salmon farm, which may contain upwards of one million fish, generates a tremendous amount of waste, largely due to a fundamental rule of animal physiology: only a small proportion of the food ingested by an animal is actually incorporated into its body as flesh, organs, and other components; the rest is used for basic life support and discharged as feces or urine. If the fish farm is located in an area with strong tides or currents, much of this waste will be dispersed and diluted. But if the flow rate through the bay is low, the waste products accumulate and can cause algal blooms, dead zones, and other environmental problems.

A second concern has to do with the efficiency of salmon farm-ing. Salmon, unlike cows or sheep, are predators—they eat other fish. To maintain salmon in captivity, farmers feed them a fish chow made up of smaller, less valuable fish that have been harvested from the ocean. And, for the same physiological reason noted above, it requires far more than a pound of fish chow to produce a pound of salmon. A study in 2000 estimated that more than three pounds of wild fish are required to produce a single pound of farmed salmon.[20] Given that there are salmon farms in cold waters around the world, from Norway to New Zealand, and these farms collec-tively raise millions upon millions of salmon, the amount of fish that are being captured, ground up, and fed to the captive salmon must be staggering. How the redirection of all this marine life into salmon farms is affecting other components of the ocean ecosystem is anyone's guess.

Although the best thing for the wild salmon of Maine would surely be an end to salmon farming in Cobscook Bay and else-where, there is little chance of that happening. To the contrary, aquaculture seems destined to grow in the years ahead as the worldwide demand for seafood grows. A more realistic goal, there-fore, may be to continue to prod the aquaculture industry to make its practices as environmentally benign as possible. In recent years, the U.S. Fish and Wildlife Service, National Marine Fisheries Ser-vice, and Army Corps of Engineers have worked with the industry to eliminate the use of foreign strains of Atlantic salmon in fish farms, to improve containment measures, and to require the mark-ing of pen-raised salmon to better track escapees.

Although the current status of wild Atlantic salmon in the United States lies somewhere between dire and dismal, there is nonetheless some reason to be hopeful. In the fall of 2006, the State of Maine opened a brief recreational fishing season for Atlantic salmon along the Penobscot River. It was strictly a catch-and-release operation, and no barbed hooks were allowed. State fisheries biologists admit the decision to open a fishing season was based less on evidence of a rebounding salmon population than on a desire to win over the hearts, minds, and wallets of anglers. But it nonetheless reflects a cautious optimism that we have finally turned the corner in terms of saving Maine's wild salmon.

Elsewhere in New England, efforts are under way to restore salmon to rivers where they once occurred but have long since vanished. Since the late 1960s, for example, fisheries managers from the federal government and the states of New Hampshire, Massachusetts, Connecticut, and Vermont have been stocking millions of Atlantic salmon fry in the Connecticut River basin, where salmon have been missing for over a century. The return on this investment has not been great—roughly two hundred to five hundred adult fish have been returning each year for the past two decades—perhaps because the habitat in this heavily settled watershed is too degraded. However, recent efforts to remove unneeded dams and to retrofit others with fishways, combined with some habitat restoration programs, may yet accomplish what the hatcheries have failed to do: produce a self-sustaining wild population of Atlantic salmon in the Connecticut River.

The United States is by no means the only nation struggling to protect its Atlantic salmon populations. In fact, the species appears to be in trouble throughout its range, according to a 2001 report by the World Wildlife Fund (also called the Worldwide Fund for Nature).[21] Over the past thirty years, for example, Canada's wild Atlantic salmon population has dropped by over 75 percent, from roughly 1.5 million fish in the 1970s to 350,000 in 2000. In Finland, wild salmon have disappeared from twenty-three of the twenty-five rivers where they historically spawned. In France, they have vanished from nearly a third of their historic spawning rivers, and they are considered at risk of extinction in the remaining two-thirds. Even in Norway, which harbors the world's largest population of wild Atlantic salmon, the fish are in trouble. Wild salmon have disappeared from about 10 percent of the streams in Norway where they once spawned and are at risk of extinction in another 33 percent. These more recent declines in European salmon populations are in addition to historical declines dating back to the Industrial Revolution.

The threats to Atlantic salmon in Canada and Europe are not very different from the ones affecting the fish in the United States—overfishing, dams, aquaculture, pollution from agriculture, industry, and cities, and so on—although their relative importance varies from country to country. A number of European nations have taken steps to reverse these declines by, among other things, reducing the allowable catch of salmon, imposing tougher regulations on fish farms and hatcheries, restoring vegetation along degraded streams, and protecting key spawning areas. Whether these actions will prove sufficient to sustain healthy wild populations of Atlantic salmon over the long term is unknown. In Europe as in North America, fisheries managers have discovered that it is far easier to protect a healthy salmon population than to rebuild a faltering one.

WE ARE IN THE PROCESS of relearning this same lesson—for the umpteenth time—in the Pacific Northwest of the United States.[22] When Atlantic salmon runs in New England and Canada were starting to fade in the early to mid-1800s, word spread of fantastic numbers of salmon on the Pacific Coast. Settlers were said to be capturing wagonloads of salmon from the spawning rivers and using them for fertilizer and pig food. To easterners who were finding it increasingly difficult (and expensive) to procure fresh salmon locally, the news from the West Coast must have seemed too good to be true. But it was true: hundreds of millions of fish belonging to seven species of salmon and seagoing trout were swimming up the major rivers from California to Alaska. Salmon exports from the West Coast to the East Coast started in 1829, but the big commercial push did not begin until several decades later, when canning technology reached the West Coast and replaced salting as the preferred method of preserving fish. In 1864, the West Coast's first commercial cannery opened on the banks of the Sacramento River.

What happened next is more or less a repeat of what happened in New England: ruthless overexploitation. Within three decades, every major river in California and Oregon was home to at least one cannery and often many more. Moreover, a growing human population began to degrade the habitat of the salmon via logging, mining, dam construction, farming, and, ultimately, urbanization. Not surprisingly, the outcome was the same: the once-great runs of salmon in Washington, Oregon, and California were reduced to a trickle. However, unlike the situation on the East Coast, where protection of the Atlantic salmon has been relatively uncontroversial (notwithstanding some hysterics around the time the Fish and Wildlife Service added Maine's salmon to the endangered species list), the conservation of Pacific salmon has been deeply contentious. The difference is largely a function of timing, ecology, and perception. First, New England was settled and developed long before the Pacific Northwest. By the time a conservation ethic developed in the United States around the end of the nineteenth century, Atlantic salmon had all but disappeared from the eastern United States. In contrast, salmon are still widespread in the Pacific Northwest; hence, there is still a lot worth fighting over. Second, many of the Pacific salmon undertake spectacular, long migrations upriver, often traveling hundreds of miles inland. The length of their journeys ensures that they conflict with a wide range of human activities. Third, the Pacific Northwest is home to some of the nation's greatest dams, built to supply the region with cheap hydropower, abundant water for irrigation, and easy access to coastal ports. Thus, the traditional salmon versus dams controversy plays out with a vengeance here.

Three statistics convey the precarious state of Pacific

salmon.[23] First, historians estimate that somewhere between 7.5 million and 16 million salmon migrated annually up the Columbia River alone prior to the arrival of white settlers. In recent years, the number of fish making that journey has ranged from roughly 580,000 to 1.2 million, the vast majority of which come from hatcheries. In fact, without hatcheries, the Columbia would have almost no salmon today. Second, a recent study by the National Research Council estimated that salmon have disappeared from about 40 percent of their historical range in California, Oregon, Washington, and Idaho over the past century. Moreover, in many of the places where the fish still occur, they are greatly reduced in numbers, as is the case with the Columbia River. Third, a 1991 report by the American Fisheries Society concluded that 106 populations ("stocks" in the parlance of fisheries biologists) of salmon in the Pacific Northwest were extinct and an additional 159 faced a moderate to high risk of extinction.

The decline of wild salmon has important ecological consequences for the Northwest.[24] Spawning salmon are an important food source for bald eagles, grizzly bears, mink, and approximately two dozen other species of birds and mammals. But salmon are much more than mere prey: as they migrate upstream, spawn, and die, they transport nutrients from the ocean to rivers and streams. A portion of these nutrients is derived from the feces, sperm, and eggs of the live fish; even more comes from the decaying carcasses of the adults, which die after spawning. These nutrients in turn boost the productivity of the spawning streams, with ramifications up and down the food chain. Phosphorus and nitrogen from salmon carcasses, for example, boost the production of phytoplankton and zooplankton, which in turn provide food for small fish, including young salmon. Through this feedback loop, salmon fry are literally sustained by their parents.

A team of scientists estimates that prior to European settlement, some 350 million to 500 million pounds of salmon migrated up the rivers of Washington, Oregon, California, and Idaho each year. Today, the amount or biomass of salmon moving up those rivers is only about 26 milliion to 30 million pounds. The scientists further estimate that the region now suffers an annual "nutrient deficit" of 11 million to 15 million pounds of marine-derived nitrogen and phosphorus. This nutrient shortfall may prove to be a serious impediment to restoring the Northwest's once-great salmon runs.[25] (The shortfall may also affect agriculture. A recent study has shown that up to a quarter of the nitrogen in grape leaves in some California vineyards is derived from salmon that have spawned, died, and decayed in nearby rivers[26]).

Reviewing the history of efforts to prevent the decline of the Northwest's salmon is a lesson in déjà vu. One sees many of the mistakes that were made on the East Coast repeated, plus a few new ones. In 1896, for example, Oregon's first fish and game protector, Hollister McGuire, issued a report outlining the steps necessary to save the state's declining salmon.[27] McGuire called on authorities to impose closed seasons at the appropriate times of the year to ensure that a sufficient number of adult salmon lived long enough to spawn. He also urged the state to prohibit fishing in the spawning streams or at their mouths, to install fishways on all dams, and to require farmers to place screens in front of their irrigation and drainage ditches to prevent young salmon from entering them during their seaward migration. And last, he called for the establishment of hatcheries to bolster declining runs.

McGuire did not list habitat destruction as a threat, perhaps because Oregon was relatively undeveloped at that time.[28] But otherwise, a number of McGuire's recommendations mirror those offered nearly thirty years earlier by Maine's commissioner of fisheries to prevent the loss of Atlantic salmon. Sadly, the only recommendations by McGuire that were followed in earnest were those pertaining to closed seasons and the construction of more hatcheries. On the West Coast as on the East Coast, political expediency and the promise of a technological solution caused politicians to reject the tough choices.

Perhaps McGuire's inattention to salmon habitat made sense at the time, but it is now abundantly clear that throughout the Northwest the world of the wild salmon has become highly degraded, polluted, and fragmented. In the Sierra Nevada of California, for example, hydraulic mining during the gold rush resulted in the deposition of tons of sediments in streams and rivers, destroying countless miles of salmon spawning habitat. Some of those streams and rivers remain unfit for salmon to this day. Logging did not have a major effect on Northwest salmon until after the Second World War, when timber companies turned to the region's magnificent old-growth forests to supply the lumber needs of a growing economy. Throughout the 1960s, 1970s, and 1980s, thousands of miles of logging roads were punched into the wilderness in order to clear thousands of square miles of virgin forest. Silt running off the roads and clear-cuts smothered spawning grounds; the cutting of trees along stream banks removed the shade that kept water temperatures cool enough for salmon; and foresters removed trees that had fallen naturally into the rivers, not realizing that those trees were creating the deeper pools the salmon liked.

The greatest change, of course, came with the construction of dams on the Northwest's major rivers. The era of dam building in this part of the United States began around the start of the twentieth century as demand for water and power grew in concert with the region's population. It peaked during the Great Depression, when the Roosevelt administration delivered on its promise to develop the Columbia River's potential for hydropower. By this time, the dam builders were well aware of the needs of salmon, and they were confident they could design dams in such a way that the migrations would not be blocked. Unfortunately, their confidence exceeded their capabilities. Most of the dams were designed to allow adult fish to get upstream, but the builders failed to consider the needs of young salmon heading downstream. These young fish are pressed for time. They are undergoing the physiological trans-formation ("smoltification") that will enable them to switch from a life in freshwater to a life in saltwater. If dam construction has turned a fast-flowing river into a string of slack-water pools, the journey downstream takes more time. This delay can be fatal to smolts, especially if they complete their metamorphosis long before they reach the ocean.

Once the dams were in place, the stage was set for an expan-sion of agriculture, since irrigation became much more feasible. This, too, redounded to the detriment of the salmon as water levels were drawn down during the crucial summer months and as silt, pesticides, and herbicides seeped off farm fields into streams and rivers.

Although state and federal officials have fretted over salmon in the Pacific Northwest for decades, the 1990s marked the beginning of a much more concerted (and controversial) effort to save remain-ing stocks. On January 3, 1992, for example, the Fish and Wildlife Service added the Snake River sockeye to the endangered species list, making it one of the first Pacific salmon stocks to receive feder-al protection. Historically, few fish traveled as far or climbed as high as the Snake River sockeyes. For countless generations, they traveled up the Columbia River and then along the Snake River and Salmon River until they reached Idaho's Redfish Lake, a one-way journey of roughly nine hundred miles. A difficult trip under any circumstance, it became an all-but-impossible obstacle course after the Great Depression, with eight enormous hydropower dams blocking the route, reduced water levels due to withdrawals for irri-gation, habitat degradation due to logging and livestock grazing, and, lurking offshore, a growing number of fishing vessels. Perhaps not surprisingly, the number of sockeyes returning to Redfish Lake to spawn each year in the early 1990s dropped to single digits. In 1992, with the power of the Endangered Species Act unleashed on

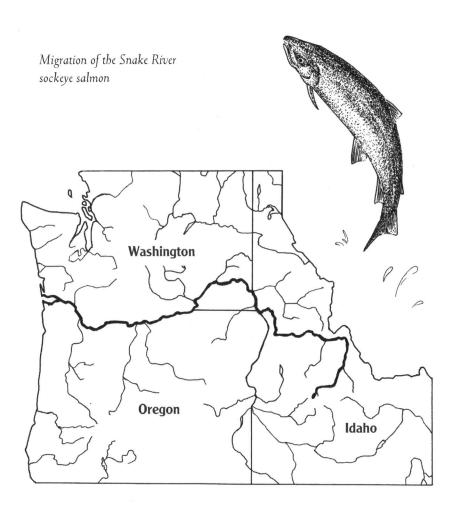

Migration of the Snake River sockeye salmon

behalf of the fish, no one—not state or federal officials, not farmers or loggers or dam operators, not anyone in the region who used electricity—could ignore the problem.

By the end of 2006, the Fish and Wildlife Service added twenty-six additional stocks or populations of Northwest salmon and seagoing trout to the endangered species list, extending the reach of the Endangered Species Act to big and small watersheds from California to the U.S.-Canadian border. To farmers and loggers worried about their livelihoods, these decisions must have seemed like symptoms of ichthyological madness. None of the Northwest's salmon is in danger of disappearing completely, so why worry about the fate of particular populations? Why add the population of sockeye that spawns in Redfish Lake to the endangered list when there are plenty of sockeye in Alaska? Why invoke the power of the Endangered Species Act to protect the chum salmon of the Columbia River or the chinooks of the upper Willamette River if there are healthy populations of chums and chinooks in other rivers? The answer is that each of these populations or stocks exhibits unique genetic, physiological, or behavioral traits that enable the fish to thrive in each river. Each salmon species is the aggregate of its constituent stocks, and those stocks are not interchangeable. This was the lesson the hatchery managers learned the hard way, and it was now being taught to all the citizens of the Pacific Northwest.

Across the Pacific Northwest, there have been battles over water allocation (essentially pitting salmon against farmers during dry years), arguments over the operation of the major dams, as well as restrictions on livestock grazing, logging, and the construction of logging roads in the national forests that harbor the spawning grounds of many salmon stocks. There have been the usual rounds of finger-pointing, with farmers blaming fishers, fishers blaming dam operators, and dam operators blaming fishers and farmers. But there has also been cooperation—not everywhere, certainly, and

not all the time, but often enough to suggest people may be growing weary of fighting and are ready to negotiate a settlement to the
region's water wars. To cite one example, in Oregon an unusual
coalition of Indian tribes, farmers, ranchers, environmentalists, and
government officials have banded together to form the Deschutes
River Conservancy.[29] The organization's goal is to secure enough
water in the ditched, dammed, and overused Deschutes to restore
its chinook, sockeye, and steelhead runs.[30]

To accomplish this goal, the conservancy intends to raise
money to improve the efficiency of irrigation practices. Currently, a
large fraction of the water intended for irrigation is lost as it seeps
into the ground underlying the ditches. Lining the ditches with
plastic pipe reduces these losses, so much so that more acres can be
irrigated with less water, leaving more water in the river for fish.
Moreover, as undesirable as residential sprawl is in so many ways,
there is a silver lining with respect to water at least: in this part of
Oregon, an acre of houses uses only about a third of the water
needed to irrigate an acre of farmland. Thus, the rapid population
growth around towns like Bend or Redmond, which results in the
conversion of farmland into housing developments, potentially
frees up water that can be used on behalf of salmon restoration.
Time will tell whether the fragile coalition embodied by the
Deschutes River Conservancy (and similar groups forming elsewhere in the region) can stay together long enough and find sufficient common ground to witness a return of migrating salmon. But
for now, the cooperation is a welcome respite from the more traditional legal and political battles that have generated a lot of acrimo-
ny but not a lot of salmon.

Paradoxically, even as wild salmon populations were declining
across Europe and North America, people were releasing Atlantic
and Pacific salmon in far-flung places around the world in order to
create new opportunities for sport fishing.[31] Today, recreational
fishers flock to the Patagonia region of Argentina and Chile where,

starting in the austral spring and lasting through the fall, introduced populations of Atlantic salmon migrate up the rivers to spawn. Springtime also marks the start of Patagonia's eagerly anticipated chinook migration; this Pacific Northwest species apparently became established here when fish escaped from a local hatchery. (Chile, it should be noted, has embraced salmon farming and is now the world's second largest exporter of farmed salmon.) In New Zealand, chinook salmon were introduced from California at the start of the twentieth century, and spawning runs are now established in a number of rivers. Fisheries managers have even tried to create populations of Atlantic, chinook, coho, and pink salmon in the Great Lakes of the United States and Canada. The fish spawn in the tributaries, and the young salmon migrate to the Great Lakes to grow (apparently having bypassed the physiological changes associated with a switch to saltwater).[32]

Given the extent to which wild salmon populations have been harmed by humans, it's tempting to cheer these efforts to create new populations in different corners of the world. But ecological considerations suggest caution. Salmon, it should be remembered, are ferocious predators. As youngsters they dine on invertebrates and little fish. As adults they eat larger fish and shrimp. One wonders what their effect will be on other species in the marine and freshwater ecosystems of Chile, Argentina, New Zealand, and elsewhere where they have been introduced.

IN AN ODD SORT OF WAY, the history of salmon parallels that of the bison. In both cases, we took species that were once abundant and free roaming, harvested them unsustainably, destroyed their habitats, and replaced them with domesticated versions of themselves (cattle on the Great Plains, farmed salmon in Maine and elsewhere). And in much the same way that dedicated individuals banded together to save the last of the bison at the end of the nineteenth century, so too are people beginning to rally around the last of the wild salmon, eager to preserve one of the world's great migratory phenomena. We may well succeed in protecting some of the remaining salmon runs in North America and Europe, and we may even be able to restore fish to a few of the rivers where they vanished long ago. But it seems increasingly clear that the only migration most salmon will make in the future is the journey from the fish farm to the canning factory.

No Way Home?

THERE MAY COME A TIME when a million truly wild salmon make their way up the Columbia River to spawn each year instead of the million hatchery-reared fish that do so today. Such a transformation would be difficult, time-consuming, and expensive to achieve. It would require habitat-restoration programs up and down the Columbia and its tributaries, restrictions on offshore fishing, and perhaps the breaching of several dams. Dozens of institutions and agencies and thousands of individuals would have to embrace the idea, and it would cost hundreds of millions of dollars. But it could be done. On the other hand, I am reasonably certain we will never again see the Columbia teeming with tens of millions of wild salmon as it did at the start of the nineteenth century—not with all the demands we now make on that river in terms of hydropower,

Bison on the Great Plains

irrigation, and navigation. Similarly, the American Prairie Foundation may well succeed in restoring a free-roaming herd of bison in
eastern Montana, but it will never re-create the herds of millions
that once spanned the Great Plains. Nor does the foundation even
aspire to do so, given the degree to which people have converted the
once-vast grasslands of the midcontinent into farmland. A similar
situation pertains to springbok in South Africa. The springbok has
not declined to the point of being in danger of extinction, but its
days as one of the world's greatest migratory spectacles are over.

History and common sense suggest that spectacular migrations
cannot be re-created once they have been destroyed. In a few cases,
we may be able to restore a semblance of them (often at great
expense and with considerable effort), but with an expanding
human population and its growing demands for natural resources,
the barriers to restoring enough acres of land or miles of rivers to
fully re-create such phenomena are virtually insurmountable. Thus,
great migrations are best viewed as irreplaceable treasures, increasingly scarce reminders of a time when humans did not dominate
the earth. Protecting the few such spectacles that remain, whether
they involve wildebeest and zebras in the Serengeti, sandhill cranes
along the Platte River, monarchs in Mexico, or caribou in the Arctic, should be a top priority for conservationists around the world.

Still other migrations are notable not for the sheer number of
animals involved, but rather as feats of endurance. Conservation
biologist Joel Berger, for example, has urged the federal government
to protect the migratory pathway of the pronghorn that annually
trek back and forth between Grand Teton National Park and the
upper Green River Basin in Wyoming, a three-hundred-mile journey that ranks as one of the longest terrestrial mammal migrations
on earth. The number of pronghorn making that journey is down
to a couple hundred, and their disappearance would have little effect
on the overall well-being of the species. But it would mark the end

of a singularly remarkable (and irreplaceable) migration. Similarly, the nine-hundred-mile pilgrimage each Snake River sockeye makes as it travels from the Pacific Ocean to Idaho's Redfish Lake—the longest migration of any remaining salmon population in North America—fully justifies the federal and state efforts now under way to save these particular fish.

The struggle to protect the Green River pronghorn and Snake River sockeyes may prove to be milestones in conservation history. People have been fascinated by animal migration for millennia, but until recently we rarely viewed it as a phenomenon with intrinsic value, at least not until the migrating species themselves teetered on the brink of extinction. Thus, although residents of the eastern United States were reporting a sharp drop in the number of migrating passenger pigeons as far back as the late 1600s, nothing much was done to protect the birds until the end of the nineteenth century, by which time it was too late. (The last passenger pigeon died in captivity in 1914.) Similarly, it was the imminent extinction of the bison, not the loss of the great herds (which occurred a few decades earlier), that finally jolted people into ending the unrestrained slaughter of this species. Today's efforts to protect monarch butterflies in Mexico, migratory songbirds wintering in Central America, and individual salmon stocks in the United States represent a subtle but important change in attitude, a dawning awareness of the intrinsic value of migratory phenomena and a desire to protect migratory species while they are still relatively common.

History provides several straightforward and important lessons regarding the challenges we face in attempting to save migratory phenomena. First, the difficulty of protecting a migration increases with the number of jurisdictions the animals cross. Protecting the Platte River's sandhill cranes is hardly a simple task—a variety of competing interests clamor for the water within the river and the land surrounding it—but the future of that remarkable migration fundamentally rests with just two nations, the United States and

Canada, and both have a long history of friendship and cooperation. Contrast that situation with the case of the red knots that fly from their wintering grounds in Tierra del Fuego to their breeding grounds in the Canadian Arctic. Protecting these birds means preserving a network of refueling sites in South and North America, plus adequate amounts of breeding and wintering habitat. A half-dozen or more countries need to work together in a coordinated fashion, and that's no easy task.

Nor is it just the length of the journey and the number of intervening countries that determine the difficulty of protecting a particular migratory phenomenon. The number and nature of the stopover sites is also critically important. Thus, the red knot may actually prove to be easier to protect than a more numerous bird like the cerulean warbler. In the case of the knots, scientists think they know where all the key refueling stops are. The question is whether we (and this "we" encompasses the citizens of those half-dozen countries where most of the knots stop to rest and refuel) have the political will to protect those sites. Cerulean warblers, on the other hand, have a much shorter journey. They breed in the eastern United States and Canada and migrate to northern South America. But individual warblers probably vary their rest and refueling sites each year in response to local weather conditions, wind velocities, and other factors that affect how far each bird is able to fly in a given night. As a result, the warblers migrate across a relatively broad front. No one site is likely to be crucial to their welfare, but some number and combination of sites are.[1] This uncertainty raises the question, where do we start in our efforts to protect such a species? And where do we stop? How do we know if we have saved enough breeding, wintering, and migratory habitat? Given that cerulean warblers are declining at an alarming rate (based on counts of birds on the nesting grounds), we can be reasonably certain that we have not protected enough habitat. But because we do not know whether the declines reflect losses of breeding, wintering, or stopover habitat, or some combination of the three, we do not know where to focus our conservation efforts. Until we come up with the answer, the cerulean warbler faces continued declines and perhaps even extinction.

Most important of all, the likelihood of preserving a migratory phenomenon depends on the degree to which people are usurping the space and resources the migrants require. Both the North Atlantic right whale and the gray whale were reduced to a handful of individuals due to overexploitation, and both received protection at roughly the same time. Yet over the past few decades, the gray whale has staged a spectacular recovery (in the eastern Pacific at least), while the North Atlantic right whale barely hangs on. The improving status of the gray whale stems in large part from the fact that it migrates along the Pacific Coast, where it has to contend with less shipping traffic and fixed-gear fishing than the North Atlantic right whale, which migrates close to shore along the heavily traveled, heavily fished East Coast of the United States.

Over the past two decades, technology has revolutionized our understanding of animal migration. It has given us the ability to track creatures as small as a dragonfly and as wide-ranging as a leatherback sea turtle. It can tell us where animals go, how long they stay, and what they do when they get there, all of which is critical for developing coherent conservation plans. But technology alone cannot address the two fundamental challenges for protecting migration: first, governments and individuals must be willing to work cooperatively across political and jurisdictional boundaries, and, second, conservation efforts must be proactive. The first rule is simply a consequence of the fact that migratory animals respect no political boundaries. The second requirement is less a matter of fact than a matter of values. One of the most inspiring aspects of migration is the sheer number of animals involved. A dozen cranes rising from the Platte River at dawn during their spring migration is a beautiful sight; five hundred thousand doing the same thing is miraculous. Protecting the abundance of migration is key to protecting the glory of migration.

Unfortunately, conservation efforts around the world tend to be fragmented and territorial, precisely the wrong way to protect phenomena that disregard political boundaries. Berger understood this reality when he called for a federal effort to identify and protect important migratory corridors within the United States. He recognized that the efforts of the individual western states to protect migratory herds of pronghorn, elk, and bison were simply not enough, as evidenced by the number of migrations that had disappeared over time. In this regard, the current plight of Yellowstone National Park's bison is especially alarming. Their migration is a relatively simple affair: as winter hits the Northern Rockies, they move from the higher elevations of Yellowstone National Park to lower-elevation areas that are privately owned, state-owned, or under the management of other federal agencies (e.g., the U.S. Forest Service). Most of these bison never leave Montana, but they do move from jurisdiction to jurisdiction within the state. The varying mandates and agendas of Montana's different agencies and landowners have created the current dispute over the (remote) possibility of bison transmitting brucellosis to cattle. The result, unfortunately, has been a hail of bullets for those bison that dare to exit the park.

Admittedly, bison are an unusual case in that they are perceived by ranchers to pose a direct threat to the cattle industry. Such conflicts do not exist in the case of most migratory species. However, if protecting an iconic migratory animal that moves wholly within one country—indeed, within one state within that country—has proved so difficult, what is the likelihood of protecting those species that migrate through multiple countries?

In fact, there are successful models of international cooperation to protect migratory species. The Serengeti ecosystem is one example. The creation of two protected areas—Serengeti National Park in Tanzania and Masai Mara National Reserve in adjacent Kenya— has safeguarded the most spectacular mammal migration left on earth, at least for the time being. The boundaries of the two reserves encompass almost all the wet- and dry-season ranges and migratory routes used by the migratory herds of wildebeest and zebras. Although it is certainly true that a shift in migratory patterns, a slackening of enforcement, or a disease epidemic could throw the whole phenomenon into peril, that two nations jointly created a reserve system that more or less corresponds to the migratory pathways of the animals should be an inspiration to conservationists around the world.

The various treaties that ended the overexploitation of the great whales are another example of successful international cooperation. These agreements were an effective response to the plight of animals that swim across national borders or dwell in international waters where no nation has sovereignty. Notwithstanding the ongoing whaling operations of Norway, Iceland, and Japan, the future of most of the great whales is far more secure today than it was just a few decades ago. The strength of the International Convention for the Regulation of Whaling, which came into force in 1948, was that it created a regulatory body (the International Whaling Commission) with the power to stop the harvest of declining species. For many years the commission was hesitant to exercise its authority, but it eventually did so; and the results, measured in terms of increasing populations of whales, are encouraging.

A third example of a successful international agreement is the North American Waterfowl Management Plan signed by Canada and the United States in 1986, with Mexico joining in 1994. It created a multinational framework for regulating hunting and protecting habitats for thirty-seven species of ducks, geese, and swans. To date, nearly sixteen million acres of waterfowl habitat have been protected or restored under the plan.[2] Its success is due to three factors. First, the plan created a strong administrative structure for managing waterfowl populations. Second, it has a stable source of funding, derived from a tax on the sale of guns and ammunition in the United States. And third, it is backed by a large and powerful constituency, namely hunters and birdwatchers.3

In contrast, the 1979 Convention on the Conservation of Migratory Species of Wild Animals, also known as the Bonn Convention, was designed to protect migratory wildlife of all kinds, but it contains neither strong, enforceable conservation measures nor a powerful administrative structure. Hence, its main accomplishments have been a handful of agreements covering a relatively small number of species. (Regrettably, there has been no groundswell of support for the notion of giving the Bonn Convention more teeth.)

Yet even a stronger treaty is unlikely to accomplish very much as long as the current international disparities in funding and legislation persist. Quite simply, richer countries have more resources—financial, legal, and educational—to devote to conservation than do poorer countries. For example, the total annual budget for the Masai Mara National Reserve and Serengeti National Park is roughly $10 million.4 In fiscal year 2004, the U.S. federal government spent approximately $17.5 million to conserve just the Snake River sockeye, one of over two dozen salmon populations on the endangered species list.5

Of an estimated $6 billion spent annually to manage protected areas worldwide, less than 12 percent of that amount is spent in less-developed countries (where, ironically, most of the earth's biodiversity resides).6 This is not simply a reflection of the lower costs of conservation in poorer versus wealthier countries; the relative shortfall in funding for protected-area management is also greater in poorer countries than in wealthier ones. In other words, poorer countries are able to meet proportionately less of their conservation needs than wealthier countries.

This lack of money and other resources greatly hinders conservation efforts on behalf of many migratory animals. In 1994, for example, the U.S. government mandated that all fishers operating in waters from North Carolina to Texas place turtle excluder devices on their nets to avoid killing endangered sea turtles. The rule was wildly unpopular with the fishers at first, but most of them eventually complied. They were able to afford the devices, and they knew that if they failed to use them, they might be prosecuted for violating the Endangered Species Act. That's simply not a realistic scenario at the present time in places like the Philippines or Indonesia, where poverty, weak environmental laws, and poor enforcement hinder efforts to protect sea turtles and other vanishing species.

In the case of migratory species, these sorts of disparities can render the conservation efforts of any one nation moot. Thus, the United States could add the cerulean warbler to its endangered species list (as some conservationists have urged), and it could do everything in its power to secure the birds' breeding habitat, but if the forests where the warblers winter in Colombia or Ecuador are destroyed, the species may vanish nonetheless.

One is forced to conclude that if the citizens of wealthier nations value "their" migratory birds, butterflies, and other animals, they must be willing to help protect them wherever they occur. This means financial and technical assistance to private and governmental organizations in other countries. In this spirit, the American Bird Conservancy, based in The Plains, Virginia, and drawing upon a largely U.S. membership, recently paid to protect a cloud forest in Colombia that provides winter habitat for cerulean warblers.[7] Similarly, individual state chapters of The Nature Conservancy have teamed with counterpart organizations in Latin America to protect the habitats of birds that breed in North America and winter in the tropics. The U.S. federal government has also funded scientific studies and conservation programs in other countries, directed toward migratory species.

Such efforts are all to the good; there simply needs to be a lot more of them. Moreover, if we consider certain migratory phenomena to be global treasures—wildebeest in the Serengeti, leatherback turtles in the Pacific Ocean, quetzals in Costa Rica—then the citizens and governments of wealthier nations ought to be willing to contribute to their conservation, even if the animals in question never occur within the borders of the donor nations. In this spirit, The Nature Conservancy, which draws most of its financial support from donors in the United States, is working closely with African conservation organizations to protect crucial migratory corridors for mammals in Africa. Among the approaches being tried are various land-lease arrangements, whereby landowning families are paid an annual fee if they agree to leave their land undeveloped, limit the number of livestock they allow to graze on their property, and protect native wildlife.[8] The Nature Conservancy hopes to give these families an economic incentive to conserve wildlife habitat as well as the means to maintain their traditional lifestyles. Such approaches to conservation will undoubtedly raise the hackles of those who view any sort of foreign aid with suspicion, but the alternative—the loss of some of the world's greatest migratory spectacles and species—seems far more objectionable.

Given the staggering number of plants and animals now facing extinction (including 12 percent of the world's bird species and 23 percent of its mammals, according to recent analyses), the notion of protecting migratory species while they are still common may seem like a luxury.[9] It's not. There is, as noted previously, the aesthetic issue: what makes many migratory phenomena so stirring is the tremendous number of animals on the move. There are also ecological considerations. Many migratory species play important ecological roles. Monarchs pollinate flowers. Warblers and other migratory birds help to keep populations of defoliating caterpillars in check, thereby protecting forests. And when they swim upstream, spawn, die, and decay, salmon transport vital nutrients from the fertile seas to the less-fertile rivers. In all these cases, the environmental benefits are proportional to the numbers of migrants on the move. Thus, by maintaining the abundance of migratory species, we are helping to maintain the health of the environments in which they occur.

What is needed is some sort of early warning system to flag migrations in trouble, a way to alert the public well before the entire species is in peril. A number of years ago, the World Conservation Union (better known by its old acronym IUCN) adopted a uniform set of criteria to determine when a species was in danger of extinction. Earlier lists of endangered species had been based on differing standards or, worse yet, no standards at all, leading to the omission of certain species that were in trouble and the inclusion of others that were not. Among the criteria adopted by the IUCN are several related to a species' rate of decline. For example, any animal that experiences a 30 percent or greater drop in its global population over a period of ten years or three generations (whichever is longer) is considered to be vulnerable. If the same species experiences a drop of 50 percent or more over the same time period, it is considered endangered; a decline of 80 percent or more places it in the critically endangered category.[10]

It should be possible to develop similar standards to classify particular populations of migratory species. Thus, if Yellowstone's bison or Mexico's monarchs were to drop by a certain percentage over a fixed period of time, they would be considered an endangered migration. A lower rate of decline over a longer period would place the migration in the less serious category of vulnerable. Given the desirability of maintaining migration as a phenomenon of abundance, the percentage thresholds used in this case should be considerably lower than those used by the IUCN to flag species at risk of extinction. Even if such a ranking system carried no regulatory "teeth," it would provide a consistent, objective way to identify migratory phenomena at risk. A progressive country might even give imperiled migrations the same type of protection it accords endangered species.

Of course, overshadowing all our efforts to protect migratory species is the specter of global climate change. It has the potential to affect the timing of animal movements, the amount and quality of food that migrants encounter during their journeys, the quality of the habitats they find when they reach their destinations, and even, in the case of sea turtles, the gender of the migrants themselves. Given the wide range of effects that *could* occur as a result of global climate change, we face a number of difficult questions. Do we develop conservation strategies that somehow take into account the likely effects of global climate change, perhaps by expanding reserves or shifting their locations to places where we suppose conditions are likely to be favorable in the future? If so, do we even know enough about the ecology of most migratory animals to make those decisions at the present time? If funding for conservation programs is tight, how do government agencies and conservation organizations decide whether to protect the habitats migratory species need now—habitats that are being destroyed at this very moment—versus those the animals may need in another fifty years' time in response to global climate change?

There are, unfortunately, no easy solutions to this dilemma, no obvious ways to balance the urgent needs of the present with what are likely to be equally urgent needs in the future. But this much seems clear: unless we come to grips with the problem of global warming, we are unlikely to be successful in saving many of the world's migrations, regardless of what we do in terms of safeguarding current breeding grounds, wintering grounds, or migratory routes. (Of course, addressing the issue of global climate change will require an unprecedented degree of international cooperation.

This raises the hopeful possibility that ongoing efforts to address global climate change will spawn agreements and partnerships that can serve as models for addressing the plight of migratory animals.)

Looking ahead, if we are successful at preserving many of the world's imperiled migrations, it will be because we have learned to cooperate across social, political, and administrative boundaries, and because we have recognized the importance of tackling global climate change and other environmental problems long before they become crises. Few lessons could be more important for the long-term welfare of migratory animals and people alike.

IT IS A PERFECTLY ROUTINE EVENT repeated billions of times every spring and fall: a bird launches itself into the night sky—perhaps a Swainson's thrush in the United States, a nightingale in Germany, a dusky warbler in China, or a white-crested elaenia in Chile. It flies upward into the darkness, joining countless other birds engaged in similar journeys. To us, that flight into the void seems like an act of faith or courage, but the birds themselves are incapable of such emotions. Instead they are driven by instinct, by an internal compulsion crafted by natural selection over the course of millions of years. That instinct tells the bird it is time to go, time to venture hundreds or thousands of miles to some other place where living conditions will be better for the next few months, as has been the case for generations upon generations of its ancestors. The bird travels without any knowledge of what may have happened to its breeding grounds, its wintering grounds, or any of the places in between since the last time it made the journey. It just goes. In that respect, perhaps migration is an act of faith after all, a hardwired belief that there is somewhere to go and a way to get back.

Notes

INTRODUCTION: ON THE MOVE

1. In "Let's Do It," Porter refers to "educated fleas," rather than bats with fleas. How ectoparasites such as fleas handle the migrations of their hosts is a fascinating and underresearched question in ecology.

2. A. Calhoun, University of Maine, pers. comm.

3. In some cases, the definition of migration can be stretched to include daily movements of animals such as krill, provided the animals in question spend blocks of time in each locale as opposed to hustling back and forth many times over the course of the day.

4. There is some controversy as to how many bison actually inhabited the Great Plains. For more information, see D. S. Wilcove (2000), *The Condor's Shadow: The Loss and Recovery of Wildlife in America* (New York: Anchor Books).

5. BirdLife International (2007), Expedition solves aquatic warbler mystery, February 22, www.birdlife.org/news/nes/2007/02/aquatic_warbler_senegal.html (accessed February 23, 2007).

CHAPTER 1: EMPTY SKIES

1. See M. Wikelski, E. M. Tarlow, A. Raim, R. H. Diehl, R. P. Larkin, and G. H. Visser (2003), Costs of migration in free-flying songbirds, *Nature* 423:704. My account of these researchers' exploits is a composite based on this paper, an interview with Wikelski, and a night I spent in the field with him and his colleagues chasing thrushes.

2. M. S. Bowlin, W. W. Cochran, and M. C. Wikelski (2005), Biotelemetry of New World thrushes during migration: Physiology, energetics and orientation in the wild, *Integrative and Comparative Biology* 45:295–304.

3. S. A. Gauthreaux Jr., J. E. Michi, and C. G. Belser (2005), The temporal and spatial structure of the atmosphere and its influence on bird migration strategies, in *Birds of Two Worlds: The Ecology and Evolution of Migration*, ed. R. Greenberg and P. P. Marra, 182–93 (Baltimore: Johns Hopkins University Press).

4. A. Feduccia (1980), *The Age of Birds* (Cambridge, MA: Harvard University Press).

5. Data on Rock Creek Park and other U.S. sites are taken from D. S. Wilcove and S. K. Robinson (1990), The impact of forest fragmentation on bird communities in eastern North America, in *Biogeography and Ecology of Forest Bird Communities*, ed. A. Keast, 319–31 (The Hague, Netherlands: SPB Academic Publishing).

6. J. Faaborg (2002), *Saving Migrant Birds: Developing Strategies for the Future* (Austin: University of Texas Press).

7. My discussion of the effects of forest fragmentation, cowbird parasitism, and nest predation on migratory songbirds is taken primarily from Faaborg (2002); and J. W. Terborgh (1989), *Where Have All the Birds Gone?* (Princeton, NJ: Princeton University Press).

8. D. S. Wilcove (1985), Nest predation in forest tracts and the decline of migratory songbirds, *Ecology* 66:1211–14.

9. H. Mayfield (1977), Brown-headed cowbird: Agent of extermination? *American Birds* 31:107–13; M. C. Brittingham and S. A. Temple (1983), Have cowbirds caused forest songbirds to decline? *BioScience* 33:31–35.

10. J. P. Hoover and S. K. Robinson (2007), Retaliatory mafia behavior by a parasitic cowbird favors host acceptance of parasitic eggs, *Proceedings of the National Academy of Sciences* 104: 4479–83.

11. It is also possible that even within heavily fragmented forests there are occasional "good years" of low parasitism and predation, perhaps occurring just often enough to allow songbird populations to persist.

12. M. Monkkonen, P. Helle, and D. Welsh (1992), Perspectives on Palaearctic and Nearctic bird migration: Comparisons and overview of life-history and ecology of migrant passerines, *Ibis* 134 (supplement): 7–13.

13. My discussion of Old World songbirds wintering in Africa is taken primarily from C. H. Fry (1992), The Moreau ecological overview, *Ibis* 134 (supplement): 3–6; Monkkonen et al. (1992); and Faaborg (2002).

14. J. W. Terborgh (1980), The conservation status of neotropical migrants: Present and future, in *Migrant Birds in the Neotropics: Ecology, Behavior, Distribution, and Conservation*, ed. A. Keast and E. S. Morton, 22 (Washington, DC: Smithsonian Institution Press).

15. Deforestation statistics compare forest cover in 2000 with presumed forest cover eight thousand years ago and are taken from World Resources Institute's EarthTrends country profiles, http://earthtrends .wri.org (accessed February 2005).

16. P. P. Marra, K. A. Hobson, and R. T. Holmes (1998), Linking winter and summer events in a migratory bird by using stable-carbon isotopes, *Science* 282:1884-86; D. R. Norris, P. P. Marra, T. K. Kyser, T. W. Sherry, and L. M. Ratcliffe (2004), Tropical winter habitat limits reproductive success on the temperate breeding grounds in a migratory bird, *Proceedings of the Royal Society of London, Series B: Biological Sciences* 271:59-64.

17. T. S. Sillett and R. T. Holmes (2002), Variation in survivorship of a migratory songbird throughout its annual cycle, *Journal of Animal Ecology* 71:296-308.

18. My discussion of bird hunting in Europe and North Africa is taken from M. N. McCulloch, G. M. Tucker, and S. R. Baillie (1992), The hunting of migratory birds in Europe: A ringing recovery analysis, *Ibis* 134 (supplement): 55-65; Environmental News Service (2005), Cyprus tries education to halt songbird slaughter, September 20, www.ens-newswire.com/ens/sep2005/2005-09-20-02.asp (accessed May 16, 2007); and H. Youth (2003), *Winged Messengers: The Decline of Birds*, Worldwatch Paper 165 (Washington, DC: Worldwatch Institute).

19. My discussion of the red knots and Delaware Bay is taken primarily from: A. J. Baker et al. (2004), Rapid population decline in red knots: Fitness consequences of decreased refueling rates and late arrival in Delaware Bay, Proceedings of the Royal Society of London, Series B: Biological Sciences 271: 875-882; R. I. G. Morrison, R. K. Ross, and L. J. Niles (2004), Declines in wintering populations of red knots in southern South America, Condor 106: 60-70; and L. J. Niles et al. (2007), Status of the red knot (Calidris canutus rufa) in the Western Hemisphere, report to the U.S. Fish and Wildlife Service (Trenton, NJ: New Jersey Department of Environmental Protection, Endangered & Nongame Species Program).

20. See P. F. Battley, T. Piersma, M. W. Dietz, S. Tang, A. Dekinga, and K. Hulsman (2000), Empirical evidence for differential organ reductions during trans-oceanic bird flight, *Proceedings of the Royal Society of London, Series B: Biological Sciences* 267:191-95. Note that this study was conducted on a different but closely related species, the great knot (*Calidris tenuirostris*).

21. A few thousand knots from this population appear to winter along the northern coast of Brazil, but the vast majority head for Patagonia and Tierra del Fuego.

22. E. A. Walls, J. Berkson, and S. A. Smith (2002), The horseshoe crab, *Limulus polyphemus*: 200 million years of existence, 100 years of study, *Reviews in Fisheries Science* 10:39–74.

23. Walls et al. (2002).

24. The decline in horseshoe crab numbers would affect the knots' food supply in another way, too. The spawning horseshoe crabs unearth previously laid eggs as they dig to lay their own eggs. These unearthed eggs are important to birds like the red knot, which relies mostly on eggs near the surface (see Morrison et al. 2004).

25. G. Winegrad, American Bird Conservancy, pers. comm.

26. U.S. Fish and Wildlife Service (2006), Endangered and threatened wildlife and plants; review of native species that are candidates or proposed for listing as endangered or threatened; annual notice of findings on resubmitted petitions; annual description of progress on listing actions, *Federal Register* 71, no. 176 (September 12): 53756–835.

27. R. I. G. Morrison et al. (2001), Declines in North American shorebird populations, *International Wader Study Group Bulletin* 94:34–38.

28. Morrison et al. (2001, 34).

29. J. P. Myers (1986), Sex and gluttony on Delaware Bay, *Natural History*, 69–76.

30. G. V. N. Powell and R. D. Bjork (2004), Habitat linkages and the conservation of tropical biodiversity as indicated by seasonal migrations of three-wattled bellbirds, *Conservation Biology* 18:500–509.

31. D. G. Wenny and D. J. Levy (1998), Directed seed dispersal by bellbirds in a tropical cloud forest, *Proceedings of the National Academy of Sciences* 95:6204–7.

32. G. V. N. Powell and R. D. Bjork (1994), Implications of altitudinal migration for conservation strategies to protect tropical biodiversity: A case study of the resplendent quetzal *Pharomacrus mocinno* at Monteverde, Costa Rica, *Bird Conservation International* 4:161–74.

33. See C. P. van Schaik, J. W. Terborgh, and S. J. Wright (1993), The phenology of tropical forests: Adaptive significance and consequences for primary consumers, *Annual Review of Ecology and Systematics* 24:353–77. Indirectly, at least, fruit-eating animals have some control over the phenology of fruit-bearing trees. For example, as a result of natural selection, trees in a given area may synchronize their fruiting so as to swamp potential fruit predators, a safety-in-numbers strategy. Even so, the availability of water and sunlight will be critical determinants of when the trees can or cannot produce fruit.

34. P. K. Strode (2003), Implications of climate change for North

American wood warblers (Parulidae), *Global Change Biology* 9:1137–44.

35. C. Both, S. Bouwhuis, C. M. Lessells, and M. E. Visser (2006), Climate change and population declines in a long-distance migratory bird, *Nature* 441:81–83.

36. Partners in Flight is well described in Faaborg (2002).

37. See Council Directive of 2 April 1979 on the Conservation of Wild Birds, http://europa.eu.int/eur-lex/en/consleg/pdf/1979/en_1979 L0409_do_001.pdf (accessed May 16, 2007).

38. C. de Klemm (1994), The problem of migratory species in international law, in *Green Globe Yearbook of International Cooperation on Environment and Development 1994*, ed. H. O. Bergesen and G. Parmann, 67–77 (Oxford: Oxford University Press); M. J. Bowman (1999), International treaties and the global protection of birds: Part II, *Journal of Environmental Law* 11:281–300.

CHAPTER 2: A MOUNTAIN OF BUTTERFLIES AND A CLOUD OF GRASSHOPPERS

1. R. W. Russell, M. L. May, K. L. Soltesz, and J. W. Fitzpatrick (1998), Massive swarm migrations of dragonflies (Odonata) in eastern North America, *American Midland Naturalist* 140:325–42.

2. Most of my discussion of patterns of dragonfly migration is taken from Russell et al. (1998). See also D. Moskowitz, J. Moskowitz, S. Moskowitz, and H. Moskowitz (2001), Notes on a large dragonfly and butterfly migration in New Jersey, *Northeastern Naturalist* 8:483–90.

3. My discussion of radio telemetry and dragonflies is based on field trips I have taken with Martin Wikelski and on M. Wikelski, D. Moskowitz, J. Adelman, J. Cochran, D. Wilcove, and M. L. May (2006), Simple rules guide dragonfly migration, *Biology Letters* 2:325–29.

4. Moreover, if green darners increase the ratio of travel days to rest days during the course of the migration, they potentially can travel much farther than four hundred miles (R. Pasquier, Environmental Defense, pers. comm.).

5. For a review of past theories about the winter home of the monarch, see L. P. Brower (1995), Understanding and misunderstanding the migration of the monarch butterfly (Nymphalidae) in North America: 1857–1995, *Journal of the Lepidopterists' Society* 49:304–85; and F. A. Urquhart (1987), *The Monarch Butterfly: International Traveler* (Chicago: Nelson-Hall). Information on the nonmigratory populations of monarchs in Florida was provided by Lincoln Brower, Sweet Brier College, pers. comm.

6. My account of the discovery of the monarch's winter home is taken from F. A. Urquhart (1976), Found at last: The monarch's winter home, *National Geographic* (August): 161–73; Urquhart (1987); and Brower (1995).

7. Today, some monarch aficionados prefer to place an adhesive disc printed with the ID number and return address on the hindwing of the butterfly rather than fold a tag over the forewing. See Monarch Watch, www.monarchwatch.org (accessed May 20, 1997).

8. Brugger quoted in Urquhart (1976, 166). Urquhart spells the last name of his collaborator as "Brugger" in the *National Geographic* article and as "Bruger" in the book, *The Monarch Butterfly: International Traveler*. A Google search reveals far more citations to Brugger than to Bruger. The actual date of discovery may have been January 2, 1975 (see Brower 1995).

9. Urquhart (1976, 156–57).

10. Lincoln Brower (pers. comm.) reports densities of fifty million butterflies per hectare in some places, which converts to slightly more than twenty million per acre.

11. W. H. Calvert and L. P. Brower (1986), The location of monarch butterfly (*Danaus plexippus* L.) overwintering colonies in Mexico in relation to topography and climate, *Journal of the Lepidopterists' Society* 40: 164–87.

12. L. P. Brower, L. S. Fink, and P. Walford (2006), Fueling the fall migration of the monarch butterfly, *Integrative and Comparative Biology* 46:1123–42.

13. Calvert and Brower (1986).

14. S. B. Malcolm, B. J. Cockrell, and L. P. Brower (1993), Spring recolonization of eastern North America by the monarch butterfly: Successive brood or single sweep migration? In *Biology and Conservation of the Monarch Butterfly*, Sciences Series no. 38, ed. S. P. Malcolm and M. P. Zalucki, 253–67 (Los Angeles: Natural History Museum of Los Angeles).

15. My discussion of monarch navigation is based primarily on V. P. Bingman and K. Cheng (2005), Mechanisms of animal global navigation: Comparative perspectives and enduring challenges, *Ethology Ecology and Evolution* 17:295–318; and H. Mouritsen and B. J. Frost (2002), Virtual migration in tethered flying monarch butterflies reveals their orientation mechanisms, *Proceedings of the National Academy of Sciences* 99:10162–66.

16. L. P. Brower (1996), Monarch butterfly orientation: Missing pieces of a magnificent puzzle, *Journal of Experimental Biology* 99:93–103.

17. Bingman and Cheng (2005).

18. W. K. Stevens (1990), Monarchs' migration: A fragile journey, *New York Times*, December 4. Other estimates for the winter range are even smaller, for example, twenty by forty miles in L. P. Brower (1988), A place in the sun, *Animal Kingdom* 91, no. 4 (July/August): 42-51.

19. L. P. Brower, G. Castilleja, A. Peralta, J. Lopez-Garcia, L. Bojorquez-Tapia, S. Diaz, D. Melgarejo, and M. Missrie (2002), Quantitative changes in forest quality in a principal overwintering area of the monarch butterfly in Mexico: 1971-1999, *Conservation Biology* 16:346-59.

20. Brower quoted in D. C. Scott (1993), Logging squeezes butterflies' winter home, *Christian Science Monitor*, April 13.

21. J. B. Anderson, and L. P. Brower (1996), Freeze-protection of overwintering monarch butterflies in Mexico: Critical role of the forest as a blanket and an umbrella, *Ecological Entomology* 21:107-16.

22. See, for example, C. K. Yoon (2003), Monarch butterflies alive and well in Mexico, *New York Times*, February 14.

23. R. Mader, Planeta.com, Monarch butterflies, www.planeta.com/ecotravel/mexico/monarchs.html (accessed May 5, 2006).

24. C. K. Yoon (2000), Monarch butterflies lose much of their wintering grounds, *New York Times*, September 12.

25. Jiménez quoted in G. Thompson (2004), Where butterflies rest, damage runs rampant, *New York Times*, June 2.

26. See Xerces Society, California monarch butterfly conservation campaign, www.xerces.org/Monarch_Butterfly_Conservation/monarch_conservation.htm (accessed May 7, 2006).

27. My discussion of the effects of genetically modified crops on monarchs is taken largely from Brower et al. (2006).

28. My account of the demise of the Rocky Mountain locust is taken largely from J. A. Lockwood (2004), *Locust: The Devastating Rise and Mysterious Disappearance of the Insect That Shaped the American Frontier* (New York: Basic Books). Anyone wanting more information on this insect should be sure to consult Lockwood's fascinating book.

29. M. Enserink (2004), Can the war on locusts be won? *Science* 306:1880-82. Additional information on visual and chemical stimuli was provided by Iain Couzin, University of Oxford, pers. comm. Of course, I am assuming that the same developmental mechanisms that apply to extant locust species applied to the now-extinct Rocky Mountain locust.

30. Lockwood (2004, 20).

31. Department of the Interior report quoted in Lockwood (2004, 78).

32. My discussion of the desert locust is based primarily on J. Roffey and J. I. Magor (2003), *Desert Locust Population Parameters*, Technical

Series AGP/DL/TS/30 (Rome: FAO); FAO (2004), *Hunger in Their Wake: Inside the Battle against the Desert Locust* (Rome: FAO); Enserink (2004); and R. A. Mittermeier, P. R. Gil, C. G. Mittermeier, T. Brooks, M. Hoffmann, W. R. Konstant, G. A. B. Da Fonseca, and R. B. Mast (2003), *Wildlife Spectacles* (Mexico: CEMEX, Agrupación Sierra Madre, Conservation International).

33. Enserink (2004).

CHAPTER 3: IN SEARCH OF GREENER PASTURES

1. My account of the mechanisms driving the Serengeti migration is taken largely from A. R. E. Sinclair (1995), Serengeti past and present, in *Serengeti II: Dynamics, Management, and Conservation of an Ecosystem*, ed. A. R. E. Sinclair and P. Arcese, 3–30 (Chicago: University of Chicago Press); and M. G. Murray (1995), Specific nutrient requirements and migration of wildebeest, in Sinclair and Arcese, *Serengeti II*, 231–56.

2. The grazing succession in the Serengeti is detailed in S. J. McNaughton (1976), Serengeti migratory wildebeest: Facilitation of energy flow by grazing, *Science* 191:92–94; S. J. McNaughton (1985), Ecology of a grazing system: The Serengeti, *Ecological Monographs* 55:259–94; A. R. E. Sinclair (1979), Dynamics of the Serengeti ecosystem: Process and pattern, in *Serengeti: Dynamics of an Ecosystem*, ed. A. R. E. Sinclair and M. Norton-Griffiths, 1–30 (Chicago: University of Chicago Press); L. Maddock (1979), The "migration" and grazing succession, in Sinclair and Norton-Griffiths, *Serengeti*, 104–29; and R. H. V. Bell (1971), A grazing ecosystem in the Serengeti, *Scientific American* 224:86–93.

3. Sinclair (1995).

4. J. Bower (1991), The pastoral Neolithic of East Africa, *Journal of World Prehistory* 5:49–82.

5. R. H. Lamprey and R. S. Reid (2004), Expansion of human settlement in Kenya's Maasai Mara: What future for pastoralism and wildlife? *Journal of Biogeography* 31:997–1032.

6. My discussion of the rinderpest epidemic is taken from Lamprey and Reid (2004); and A. Dobson (1995), The ecology and epidemiology of rinderpest virus in Serengeti and Ngorongoro Conservation Area, in Sinclair and Arcese, *Serengeti II*, 485–505.

7. S. Serneels, M. Y. Said, and E. F. Lambin (2001), Land cover changes around a major East African wildlife reserve: The Mara ecosystem (Kenya), *International Journal of Remote Sensing* 22:3397–3420.

8. Economic values of livestock, farming, and tourism are taken from Lamprey and Reid (2004).

9. C. Foley, Wildlife Conservation Society, pers. comm.

10. The population growth rate of the Maasai is estimated at 3.9 percent per year in K. Homewood et al. (2001), Long-term changes in Serengeti-Mara wildebeest and land cover: Pastoralism, population, or policies? *Proceedings of the National Academy of Sciences* 98:12544–49. One of the main conclusions of this study is that economic factors are more important in turning pastoralists into farmers than is Maasai population growth.

11. Lamprey and Reid (2004).

12. Lamprey and Reid (2004).

13. In the case of Serengeti National Park, wildlife declines in the 1970s and 1980s can be traced to budget reductions that curtailed antipoaching patrols. When the park's budget increased in the late 1980s, resulting in more guards on patrol, wildlife numbers rebounded. See R. Hilborn et al. (2006), Effective enforcement in a conservation area, *Science* 314:1266.

14. Sinclair (1995); S. Thirgood et al. (2004), Can parks protect migratory ungulates? The case of the Serengeti wildebeest, *Animal Conservation* 7:113–20.

15. For a history of Bontebok National Park, see South African National Parks, www.sanparks.org/parks/bontebok/tourism/history.php (accessed August 12, 2005).

16. The quotation is from J. Kingdon (1997), *The Kingdon Field Guide to African Mammals* (San Diego: Academic Press), 426. The attribution of the bontebok's movements around their enclosures to innate, ancestral behavior is a guess on my part.

17. Elands and black wildebeest were sighted with the springbok herds on occasion. Both species were dramatically reduced in numbers very early during the settlement of the Karoo. Therefore, we do not know to what extent they were a part of the Karoo's mammal migration (C. Roche, Wilderness Safaris, pers. comm.).

18. L. G. Green (1955), *Karoo: The Story of the Karoos of South Africa—the Great Karoo, the Little Karoo and the Far Corners of the North West Cape and Namaqualand* (Cape Town: Howard Timmins).

19. Green (1955, 37).

20. Green (1955, 38).

21. S. C. Cronwright-Schreiner (1925), *The Migratory Springbucks of South Africa (The Trekbokke), Also an Essay on the Ostrich and a Letter Descriptive of the Zambesi Falls* (London: T. Fisher), 48–49.

22. J. D. Skinner (1993), Springbok (*Antidorcas marsupialis*) treks, *Transactions of the Royal Society of South Africa* 48:291–305.

23. C. Roche (2005), 'The springbok . . . drink the rain's blood': Indigenous knowledge and its use in environmental history—the case of the /Xam and an understanding of springbok treks, *South African Historical Journal* 53:1–22; C. Roche, South Africa's Serengeti: The ecology and history of Karoo springbok treks and their modern day potential, www.namakaroo.org/springbok_treks.htm (accessed April 11, 2007).

24. Roche (2005).

25. Green (1955, 44).

26. C. J. Roche (2004), Ornaments of the desert: Springbok treks in the Cape Colony, 1774–1908, Master of Arts in historical studies thesis, University of Cape Town.

27. Roche (2004).

28. My discussion of the white-eared kob is taken largely from J. M. Fryxell and A. R. E. Sinclair (1988), Seasonal migration by white-eared kob in relation to resources, *African Journal of Zoology* 26:17–31; and W. Knocker (2003), Why big mammals struggle to survive, *Sudan Mirror* 1, no. 1 (October 6–18), www.sudanmirror.com/archives/vol1-issue1/spc-mammals.html (accessed August 19, 2005).

29. My discussion of mammal migrations in Tarangire is based on conversations with Charles Foley and on C. Foley and L. Foley (no date), *Protecting the Tarangire-Simanjiro Dispersal Areas* (report of the WCS-Tarangire Elephant Project).

30. Other notable migrants include Tarangire National Park's elephants. Elephants, too, must exit the park during the wet season to find phosphorus-rich forage, with some herds traveling about fifty miles. In fact, there are three somewhat autonomous subpopulations of elephants residing within Tarangire, and each of these subpopulations heads to a different area when it leaves the park (Foley, pers. comm.).

31. Foley and Foley (no date, 20).

CHAPTER 4: WHERE THE BUFFALO ROAMED

1. Separating an ancient bison wallow from a depression produced by other forces (i.e., local geological processes) is apparently quite difficult. Hence, not all supposed wallows may be correctly identified, even by an experienced naturalist. See B. R. Coppedge, S. D. Fuhlendorf, D. M. Engle, B. J. Carter, and J. H. Shaw (1999), Grassland soil depressions: Relict bison wallows or inherent landscape heterogeneity? *American Midland Naturalist* 142:382–92; and D. F. Lott (2002), *American Bison: A Natural History* (Berkeley: University of California Press).

2. An unusually harsh winter might force the animals to recongregate

as they sought shelter or food in a relatively small number of sheltered or accessible places. See D. B. Bamforth (1987), Historical documents on the Great Plains, *Plains Anthropologist* 32:1–16.

3. E. T. Seton (1909), *Life-Histories of Northern Animals: An Account of the Mammals of Manitoba*, vol. 1, *Grass Eaters* (New York: Charles Scribner's Sons), 266–67.

4. Seton (1909).

5. A. D. Whicker and J. K. Detling (1988), Ecological consequences of prairie dog disturbances, *BioScience* 38:778–85. Both the white-tailed and Gunnison prairie dogs probably numbered in the tens of millions each. The Utah prairie dog, by contrast, has always been a comparatively rare and localized species. As far back as 1920, its total population was probably less than one hundred thousand.

6. Whicker and Detling (1988); K. Krueger (1986), Feeding relationships among bison, pronghorn, and prairie dogs: An experimental analysis, *Ecology* 67:760–70.

7. The term *forb* refers to any flowering plant with a woody stem that is not a grass.

8. K. F. Higgins (1986), *Interpretation and Compendium of Historical Fire Accounts in the Northern Great Plains*, Resource Publication no. 161 (Washington, DC: U.S. Fish and Wildlife Service).

9. R. E. England and A. DeVos (1969), Influence of animals on pristine conditions on the Canadian grasslands, *Journal of Range Management* 22:87–94. The authors argue that moderate to heavy grazing by bison may be beneficial to pronghorn.

10. H. R. Lamar, ed. (1998), *The New Encyclopedia of the American West* (New Haven, CT: Yale University Press).

11. A. C. Isenberg (2000), *The Destruction of the Bison: An Environmental History, 1750–1920* (Cambridge: Cambridge University Press).

12. Isenberg (2000, 83).

13. My discussion of the destruction of the bison herds is taken primarily from Isenberg (2000) and Lamar (1998).

14. Isenberg (2000, 132).

15. Delano quoted in Isenberg (2000, 152).

16. Roosevelt quoted in P. R. Cutright (1956), *Theodore Roosevelt— The Naturalist* (New York: Harper & Brothers), 41.

17. Isenberg (2000).

18. C. Freese et al. (in press), Second chance for the plains bison, *Biological Conservation*; C. Freese (2005), American prairie restoration project: Second chance for the plains bison (unpublished report).

19. Isenberg (2000); Freese (2005).

20. A case can be made that the bison herd that migrates every winter from Grand Teton National Park to the adjacent National Elk Refuge also meets the criteria of being genetically pure and free roaming. These animals migrate up to thirty miles each way. However, I am reluctant to put them on a par with the Yellowstone bison because the National Elk Refuge provides food for its wintering elk herd, which the bison also eat. In that respect, this bison herd's movement is a somewhat unnatural migration (although there are certainly unnatural aspects to the Yellowstone migration as well, such as the hazing they receive from state and federal officials who want them to head back into the park).

21. My discussion of brucellosis and its effects on Yellowstone's bison is taken largely from Lott (2002); and National Research Council (1998), *Brucellosis in the Greater Yellowstone Area* (Washington, DC: National Academies Press).

22. U.S. Department of the Interior and U.S. Department of Agriculture (2001), Notice of record of decision for final environmental impact statement and bison management plan for the State of Montana and Yellowstone National Park, *Federal Register* 66, no. 14 (January 22): 6665–66.

23. L. Bienen and G. Tabor (in press), Applying an ecosystem approach to brucellosis control: Can an old conflict between wildlife and agriculture be successfully managed? *Frontiers in Ecology and the Environment.*

24. See U.S. Department of Agriculture, Animal and Plant Health Inspection Service, www.aphis.usda.gov/vs/nahps/brucellosis (accessed November 2005).

25. Bienen and Tabor (in press).

26. J. Berger (2004) The last mile: How to sustain long-distance migration in mammals, *Conservation Biology* 18:320–31.

27. More specifically, this pronghorn migration is probably the longest mammal migration that exists anywhere between Tierra del Fuego and central Canada (J. Berger, Wildlife Conservation Society, pers. comm.).

28. D. E. Popper and F. J. Popper (1987), The Great Plains: From dust to dust, *Planning* 53 (December): 12–18.

29. Popper and Popper (1987, 17).

30. Hayden quoted in S. Charton (2004), Former Kansas governor makes startling reversal in debate about Plains' future: "I was wrong," *Kansas City Star*, March 19.

31. C. Freese, World Wildlife Fund, pers. comm.

32. According to Curt Freese (pers. comm.), "The bison made us wait, in suspense, about twenty minutes, which I reasoned wasn't asking too much of us after we made the bison wait 120 years to come back to this land."

CHAPTER 5: LOST AT SEA

1. Close relatives of the North Atlantic right whale include the North Pacific right whale (highly endangered and found primarily off the coast of Japan and Kamchatka) and the southern right whale (found in the Southern Hemisphere between 20° south and 55° south).

2. About a third of the adult North Atlantic right whales do not spend the late summer around the Bay of Fundy. Their whereabouts at that time of year remain a mystery (S. Kraus, New England Aquarium, pers. comm.).

3. My account of the history of whaling as it affects northern right whales is based largely on R. R. Reeves, T. D. Smith, and E. A. Josephson (2007), Near-annihilation of a species: Right whaling in the North Atlantic, in *The Urban Whale: North Atlantic Right Whales at the Cross-roads*, ed. S. D. Kraus and R. M. Rolland, 39–74 (Cambridge, MA: Harvard University Press); and A. Aguilar (1986), A review of old Basque whaling and its effect on the right whales (*Eubalaena glacialis*) of the North Atlantic, *Report of the International Whaling Commission*, special issue 10:191–99.

4. Reeves and colleagues (2007, 42) state that "there must have been very few, possibly only tens."

5. S. Martin (2001), *The Whales' Journey* (Crows Nest, Australia: Allen & Unwin).

6. National Marine Fisheries Service (2005), *Recovery Plan for the North Atlantic Right Whale* (Eubalaena glacialis) (Silver Spring, MD: National Marine Fisheries Service).

7. Demographic information is taken from National Marine Fisheries Service (2005).

8. S. D. Kraus et al. (2005), North Atlantic right whales in crisis, *Science* 309:561–62.

9. S. D. Kraus and R. M. Rolland (2007), The urban whale syndrome, in Kraus and Rolland, *The Urban Whale*, 488–513.

10. Kraus et al. (2005).

11. Current conservation measures for the northern right whale are taken from National Marine Fisheries Service (2005).

12. S. Kraus and M. Brown, New England Aquarium, pers. comm.

13. Kraus, pers. comm.

14. Kraus et al. (2005).

15. See S. Leatherwood and R. R. Reeves (1983), *The Sierra Club Handbook of Whales and Dolphins* (San Francisco: Sierra Club Books).

16. V. P. Bingman and K. Cheng (2005), Mechanisms of animal global navigation: Comparative perspectives and enduring challenges, *Ethology Ecology and Evolution* 17:295–318.

17. Captain C. A. Larsen before the British Interdepartmental Committee on Research and Development in the Dependencies, quoted in Martin (2001, 70).

18. My discussion of the history of the IWC is taken primarily from Martin (2001) and D. Russell (2001), *Eye of the Whale: Epic Passage from Baja to Siberia* (Washington, DC: Island Press).

19. See W. Aron, W. Burke, and M. Freeman (2002), Scientists versus whaling: Science, advocacy, and errors of judgment, *BioScience* 52:1137–40; P. J. Clapham et al. (2003), Whaling as science, *BioScience* 53:210–12; and G. Orians et al. (2003), "Scientists versus Whaling": Whose errors of judgment? *BioScience* 53:201–3.

20. K. Kaschner and D. Pauly (2004), *Competition between Marine Mammals and Fisheries: Food for Thought* (Washington, DC: Humane Society of the United States). This study does not address the issue of indirect competition between whales and humans: do whales consume krill and other small marine organisms that would otherwise be consumed by commercially important fish species? If so, then whales may be indirectly suppressing populations of fish. However, as Kaschner and Pauly point out, the major commercial fisheries are generally found along continental shelves or close to shore, places that are not prime habitats for whales, thus reducing the likelihood that whales are, indeed, important indirect competitors of humans.

21. Population trends are taken from the IWC Web site, www.iwc office.org/conservation/estimate.htm (accessed July 12, 2006).

22. S. R. Palumbi and F. Cipriano (1998), Species identification using genetic tools: The value of nuclear and mitochondrial gene sequences in whale conservation, *Journal of Heredity* 89:459–64.

23. My account of the gray whale's migration, its brush with extinction, and its subsequent recovery is taken largely from Russell (2001) and from J. Urbán, L. Rojas-Bracho, H. Pérez-Cortés, A. Gómez-Gallardo, S. L. Swartz, S. Ludwig, and R. L. Brownell Jr. (2003), A review of gray whales (*Eschrichtius robustus*) on their wintering grounds in Mexican waters, *Journal of Cetacean Research and Management* 5:281–95.

24. Urbán et al. (2003)

25. I am using a conversion factor of thirty-five barrels of oil per gray whale (Urbán et al. 2003); the details of Scammon's voyage are from Russell (2001, 164–68).

26. An effort in the late 1990s to build an evaporative salt production plant at Laguna San Ignacio generated a great deal of controversy. A number of whale biologists and environmentalists expressed concern that the nonstop pumping of seawater out of the lagoon (in order to produce salt in the evaporation ponds), the construction of piers and other infrastructure, and the resultant dredging and ship traffic would prove harmful to the whales. The plant was rejected by the Mexican government in 2000, largely due to concerns that it would harm the biosphere reserve.

27. For details on other threats to gray whales (e.g., entanglement in fishing gear, collisions with ships, and noise pollution), see Urbán et al. (2003).

28. To be precise, only the eastern Pacific population of gray whales was declared recovered. The small and struggling population in the western Pacific remains on the endangered list.

29. Data on the number of whale-watchers and their expenditures are taken from E. Hoyt (2001), *Whale Watching 2001: Worldwide Tourism Numbers, Expenditures, and Expanding Socioeconomic Benefits* (Yarmouth Port, MA: International Fund for Animal Welfare).

30. Data on the number of tourists and their expenditures in Baja is from Hoyt (2001).

31. B. Jenks, Rare, pers. comm.

32. Hoyt (2001).

33. B. W. Bowen, F. A. Abreu-Grobois, G. H. Balazs, N. Kamezaki, C. J. Limpus, and R. J. Ferl (1995), Trans-Pacific migrations of the loggerhead turtle (*Caretta caretta*) demonstrated with mitochondrial DNA markers, *Proceedings of the National Academy of Sciences* 92:3731–34.

34. K. J. Lohmann, C. M. F. Lohmann, L. M. Ehrhart, D. A. Bagley, and T. Swing (2004), Geomagnetic map used in sea-turtle navigation, *Nature* 428:909.

35. Information on the postbreeding wanderings of sea turtles is from J. R. Spotila (2004), *Sea Turtles: A Complete Guide to Their Biology, Behavior, and Conservation* (Baltimore: Johns Hopkins University Press).

36. Spotila (2004).

37. P. Matthiessen (1987), *Wildlife in America*, rev. and updated ed. (New York: Penguin Books).

38. K. Shanker (2004), Marine turtle status and conservation in the

Indian Ocean, in *Papers Presented at the Expert Consultation on Interactions between Sea Turtles and Fisheries within an Ecosystem Context*, FAO Fisheries Report no. 738 (supplement), 85–134 (Rome: FAO).

39. L. Crowder quoted in BBC News Online (2004), Longlines imperil Pacific turtles, February 13, http://news.bbc.co.uk/2/hi/science/nature/3485195.stm (accessed May 16, 2007).

40. For a discussion of turtle excluder devices, see D. S. Wilcove (2000), *The Condor's Shadow: The Loss and Recovery of Wildlife in America* (New York: Anchor Books).

41. See D. Shiode and T. Tokai (2004), A review of development, modification and implementation of TED (turtle excluder device) to reduce sea turtle bycatch in trawl fisheries, in *Papers Presented at the Expert Consultation on Interactions between Sea Turtles and Fisheries*, 171–78.

42. In one study from South Carolina, 80 percent of all loggerhead nests on two barrier islands were destroyed by raccoons (Spotila 2004).

43. L. McClenachan, J. B. C. Jackson, and M. J. H. Newman (2006), Conservation implications of historic sea turtle nesting beach loss, *Frontiers in Ecology and the Environment* 4:290–96.

44. J. B. C. Jackson (1997), Reefs since Columbus, *Coral Reefs* 16 (supplement): 23–32.

45. Spotila (2004).

46. The current population of Pacific leatherbacks is from L. Crowder (2000), Leatherback's survival will depend on an international effort, *Nature* 405:881. See also J. R. Spotila, R. D. Reina, A. C. Steyermark, P. T. Plotkin, and F. V. Paladino (2000), Pacific leatherback turtles face extinction, *Nature* 405:529–30. Current populations of green turtles and hawksbills in the Caribbean are from McClenachan et al. (2006).

47. G. C. Hays (2004), Good news for sea turtles, *Trends in Ecology and Evolution* 19:349–51.

48. K. A. Bjorndal, J. A. Wetherall, A. B. Bolten, and J. A. Mortimer (1999), Twenty-six years of green turtle nesting at Tortuguero, Costa Rica: An encouraging trend, *Conservation Biology* 13:126–34; S. Troeng and E. Rankin (2005), Long-term conservation efforts contribute to positive green turtle *Chelonia mydas* nesting trend at Tortuguero, Costa Rica, *Biological Conservation* 121:111–16.

49. My discussion of trends in the population of Kemp's ridleys and the controversial head-start program is taken largely from Spotila (2004); B. W. Bowen, T. A. Conant, and S. R. Hopkins-Murphy (1993), Where are they now? The Kemp's ridley headstart project, *Conservation*

Biology 8:853–56; and W. Holtcamp (2007), Surf's up for these sea turtles, *Defenders* 82, no. 1: 17–21.

50. In 1989, biologists discontinued the practice of bringing the
hatchlings back into captivity after first exposing them to the surf. Now
the turtles are allowed to head into the ocean after hatching. Apparently,
the scientists concluded that the additional time in captivity was not necessary to reestablish a wild population in Texas. See Holtcamp (2007).

51. Ehrenfeld quoted in Bjorndal et al. (1999, 131–32).

52. M. Oppenheimer, Princeton University, pers. comm.

53. Spotila (2004).

CHAPTER 6: AGAINST THE FLOW

1. Population tallies for 2006 are from the Maine Atlantic Salmon Commission, www.maine.gov/asc/research/trap_count_stats.shtml (accessed
December 17, 2006). For anyone keen to spot a truly wild salmon in
Maine, the task is made even more difficult because virtually all of the
state's rivers with wild salmon also have hatcheries. In other words, a fish
that looks like a wild salmon may in fact be a hatchery-reared one (C.
Bohlen, Trout Unlimited, pers. comm.).

2. Information on the ecology of Maine's endangered Atlantic salmon
plus data on the growth of the salmon farming industry are taken from
National Marine Fisheries Service and U.S. Fish and Wildlife Service
(2004), *Draft Recovery Plan for the Gulf of Maine Distinct Population Segment of Atlantic Salmon* (Salmo salar) (Silver Spring, MD: National Marine
Fisheries Service); and National Research Council (2004), *Atlantic Salmon
in Maine* (Washington, DC: National Academies Press).

3. D. R. Montgomery (2003), *King of Fish: The Thousand-Year Run of
Salmon* (Cambridge, MA: Westview Press).

4. V. P. Bingman and K. Cheng (2005), Mechanisms of animal global
navigation: Comparative perspectives and enduring challenges, *Ethology
Ecology and Evolution* 17:295–318.

5. Montgomery (2003).

6. National Marine Fisheries Service and U.S. Fish and Wildlife Service (2004).

7. Bohlen, pers. comm.

8. National Research Council (2004).

9. Bohlen, pers. comm.

10. My discussion of the effects of acid rain on salmon is taken largely
from National Research Council (2004). For insights on the history of

efforts to control acid rain in Canada and the United States, I am grateful to Lars Hedin, Princeton University.

11. Montgomery (2003).

12. Montgomery (2003).

13. National Research Council (2004, 8).

14. Bohlen, pers. comm.

15. See American Museum of Natural History, BioBulletin, Setting rivers free: When should dams come down? www.sciencebulletins.amnh. org/biobulletin/biobulletin/story1204.html; and Natural Resources Council of Maine, www.nrcm.org/issue_edwardsdam.asp (both accessed May 9, 2007).

16. Penobscot River Restoration Trust, www.penobscotriver.org/content/4030/Unprecedented_Collaboration (accessed December 24, 2006).

17. National Research Council (2004).

18. National Marine Fisheries Service and U.S. Fish and Wildlife Service (2004).

19. B. Daley (2001), Escaped farm salmon raise alarm in Maine, *Boston Globe*, February 23.

20. R. L. Naylor et al. (2000), Effect of aquaculture on world fish supplies, *Nature* 405:1017–24.

21. World Wildlife Fund (2001), *The Status of Wild Atlantic Salmon: A River by River Assessment* (Washington, DC: World Wildlife Fund).

22. Much of my discussion of the history of salmon in the Pacific Northwest is taken from Montgomery (2003).

23. Statistics on salmon declines in the Pacific Northwest are from M. Reisner (1993), *Cadillac Desert: The American West and Its Disappearing Water* (New York: Penguin Books); National Research Council (1996), *Upstream: Salmon and Society in the Pacific Northwest* (Washington, DC: National Academy Press); R. White (1995), *The Organic Machine: The Remaking of the Columbia River* (New York: Hill and Wang); J. M. Volkman (1997), *A River in Common: The Columbia River, the Salmon Ecosystem, and Water Policy*, report to the Western Water Policy Review Advisory Commission (Springfield, VA: U.S. Department of Commerce National Technical Information Service); and W. Nehlsen, J. E. Williams, and J. A. Lichatowich (1991), Pacific salmon at the crossroads: Stocks at risk from California, Oregon, Idaho, and Washington, *Fisheries* 16:4–21.

24. C. J. Cederholm, M. D. Kunze, T. Murota, and A. Sibatani (1999), Pacific salmon carcasses: Essential contributions of nutrients and energy for aquatic and terrestrial ecosystems, *Fisheries* 24:6–15.

25. T. Gresh, J. Lichatowich, and P. Schoonmaker (2000), An estima-

tion of historic and current levels of salmon production in the northeast Pacific ecosystem: Evidence of a nutrient deficit in the freshwater systems of the Pacific Northwest, *Fisheries* 25:15–21. The authors presented their data in kilograms; I have converted their figures to pounds and rounded them off.

26. J. E. Merz and P. B. Moyle (2006), Salmon, wildlife, and wine: Marine-derived nutrients in human-dominated ecosystems of central California, *Ecological Applications* 16:999–1009.

27. My account of McGuire's report and the response to it is taken from Montgomery (2003).

28. Montgomery (2003).

29. M. Jenkins (2006), A river once more, *High Country News*, October 16.

30. Steelhead are rainbow trout that behave like salmon: they are born in freshwater, migrate to the ocean to grow, and return to freshwater to spawn.

31. For information on salmon in Patagonia, New Zealand, and the Great Lakes, see Patagonia Adventures, www.argentinachileflyfishing.com; Fish and Game New Zealand, www.fishandgame.org.nz/SITE_Default/SITE_info/Fact_Sheet/chinook_salmond.asp; and University of Wisconsin Sea Grant Institute, www.seagrant.wisc.edu/greatlakesfish.

32. Although salmon appear to be holding their own in a small number of Great Lakes tributaries, by and large the fish are maintained by annual stocking programs.

CONCLUSION: NO WAY HOME?

1. The highlands of Belize may harbor significant numbers of cerulean warblers during migration, as do some mountain ridges in the southeastern United States, but whether these areas represent bona fide migratory concentration sites or simply places where these treetop-loving birds are easier to spot is unclear. See P. B. Hamel (2000), *Cerulean Warbler Status Assessment* (Fort Snelling, MN: U.S. Fish and Wildlife Service).

2. U.S. Fish and Wildlife Service, North American Waterfowl Management Plan, www.fws.gov/birdhabitat/NAWMP/index.shtm (accessed May 17, 2007).

3. C. de Klemm (1994), The problem of migratory species in international law, in *Green Globe Yearbook of International Cooperation on Environment and Development 1994*, ed. H. O. Bergesen and G. Parmann, 67–77 (Oxford: Oxford University Press).

4. See P. Musa (2004), Mushrooming lodges, sedentary Maasai: Mara

at the crossroads of history, *The East African*, May 17, www.nationmedia. com/eastafrican/17052004/Features/PART2170520043.html (accessed March 9, 2007); and United Nations Environment Program (UNEP) (2006), *Wildlife Watching and Tourism: A Study on the Benefits and Risks of a Fast Growing Tourism Activity and Its Impacts on Species* (Bonn, Germany: UNEP).

5. U.S. Fish and Wildlife Service (2004), *Federal and State Threatened and Endangered Species Expenditures: Fiscal Year 2004* (Washington, DC: U.S. Fish and Wildlife Service).

6. A. Balmford, K. J. Gaston, S. Blyth, A. James, and V. Kapos (2003), Global variation in terrestrial conservation costs, conservation benefits, and unmet conservation needs, *Proceedings of the National Academy of Sciences* 100:1046-50.

7. American Bird Conservancy, www.abcbirds.org/international/ cerw.htm (accessed January 29, 2007).

8. T. Dunkel (2007), Into Africa: Forging new partnerships to protect an ancient continent, *Nature Conservancy* 57, no. 1: 20-35.

9. Statistics on percentage of endangered birds and mammals from the IUCN Red List of threatened species, Summary statistics, www.iucn redlist.org/info/tables/table1 (accessed January 28, 2007).

10. For details on this ranking system, see the IUCN Red List of threatened species, 2001 categories and criteria (v. 3.1), www.iucnredlist. org/info/categories_criteria2001#categories (accessed January 27, 2007).

Acknowledgments

In the time it has taken me to write this book, an Arctic tern is likely to have migrated approximately eighty-eight thousand miles, the result of four round-trip journeys from the Arctic to the Antarctic. Metaphorically, at least, it's been a long journey for me too, and one that I could not have completed without the help of many friends and colleagues.

I am especially grateful to the researchers who allowed me to accompany them in the field to learn more about various aspects of animal migration. These people include Lew Ehrhart (sea turtles in Florida), Charles and Lara Foley (mammals in East Africa), Mace Hack (sandhill cranes along the Platte River), Chris Slay (North Atlantic right whales in the Bay of Fundy), Paul Strode (songbirds in Illinois), and Martin Wikelski (songbirds and dragonflies in Illinois and New Jersey, respectively). The time I spent with them was the highlight of this whole endeavor. I am also grateful to Mac Hunter and Aram Calhoun for joining me in search of migrating wildebeest in Tanzania and North Atlantic right whales in the Bay of Fundy; it is hard to imagine two finer traveling companions.

Several people graciously assisted me in researching various topics for this book. They include Becky Barak, Christine Chang, Elissa Furutani, Joon Lee, and David Marvin. Geraldine Rhodes patiently tracked down books and articles for me and handled countless logistical details. The staff of Stokes Library and Firestone Library at Princeton University demonstrated an uncanny ability to find even the most obscure books and articles. To all these people, I am very grateful.

I am also fortunate that so many colleagues were willing to take time away from their busy schedules to comment on portions of the manuscript, including Joel Berger, Curtis Bohlen, Lincoln Brower, Moira Brown, Larry Crowder, Eric Dinerstein, Andy Dobson, John Faaborg, Charles Foley, Curt Freese, Michaela Hau, Scott Kraus, Jeffrey Lockwood, Peter Moyle, Scott Robinson, Chris Roche, and Chris Slay. Roger Pasquier and Mac Hunter were kind enough to review the entire manuscript. Patrick Huang corrected my misconceptions about physics (important to orientation and telemetry) and shared his knowledge of migration. Collectively, these individuals have corrected misstatements, identified major and minor omissions, directed me to important references, repaired my prose, and freely shared their own insights and ideas. I am deeply in their debt.

Throughout this project, I had the good fortune to work with Jonathan Cobb, my editor at Island Press. Jonathan has now edited two of my books, and in both cases he has kept me on track and on message with remarkable patience and good cheer. My agent, Peter McGuigan, was instrumental in launching this project. Financial support came from Island Press and Princeton University.

As I write these words, migratory songbirds are beginning to return to the woods around my house. Their arrival this spring seems even more wondrous than in previous years, perhaps because I now have a better understanding of the challenges they face in getting here.

David S. Wilcove
Princeton, New Jersey
April 20, 2007

Index

About the Author

David S. Wilcove is professor of ecology, evolutionary biology, and public affairs at Princeton University. One of the world's leading experts on wildlife conservation, he is the author of *The Condor's Shadow: The Loss and Recovery of Wildlife in America* (Freeman 1999; Anchor Books 2000) and numerous scientific and popular articles.

About Island Press

Island Press is the only nonprofit organization in the United States whose principal purpose is the publication of books on environmental issues and natural resource management. We provide solutions-oriented information to professionals, public officials, business and community leaders, and concerned citizens who are shaping responses to environmental problems.

Since 1984, Island Press has been the leading provider of timely and practical books that take a multidisciplinary approach to critical environmental concerns. Our growing list of titles reflects our commitment to bringing the best of an expanding body of literature to the environmental community throughout North America and the world.

Support for Island Press is provided by the Agua Fund, The Geraldine R. Dodge Foundation, Doris Duke Charitable Foundation, The Ford Foundation, The William and Flora Hewlett Foundation, The Joyce Foundation, Kendeda Sustainability Fund of the Tides Foundation, The Forrest & Frances Lattner Foundation, The Henry Luce Foundation, The John D. and Catherine T. MacArthur Foundation, The Marisla Foundation, The Andrew W. Mellon Foundation, Gordon and Betty Moore Foundation, The Curtis and Edith Munson Foundation, Oak Foundation, The Overbrook Foundation, The David and Lucile Packard Foundation, Wallace Global Fund, The Winslow Foundation, and other generous donors.

The opinions expressed in this book are those of the author(s) and do not necessarily reflect the views of these foundations.